PEARSON ALWAYS LEARNING

Damian M. Lyons •
Christina Papadakis-Kanaris •
Gary M. Weiss • Arthur G. Werschulz

Fundamentals of Discrete Structures

Second Edition

Fordham University
Department of Computer and Information Sciences
New York, New York

Cover Art: Courtesy of EyeWire/Getty Images.

Pearson Learning Solutions, 501 Boylston Street, Suite 900, Boston, MA 02116
A Pearson Education Company
www.pearsoned.com

Printed in the United States of America

6 7 8 9 10 V0CR 17 16 15 14

000200010270800759

KH

ISBN 10: 1-256-38921-8
ISBN 13: 978-1-256-38921-7

To the collections in my life that give it both meaning and purpose,
and to the collection of those whom I owe so much,
even if some only remain as collections of memories.

—CPK

do mo ṁuintir a ḃeaṫaiġ mé, seolaim ṫú, mo scríḃinn.

— DML

To my son,
for whom I think time
is moving too fast,
and for my wife,
for sharing the journey with me.

—GMW

ב״ה
לזכר נשמת אמי מורתי לאה בת חיים הלוי וקלארא ז״ל
To the memory of my mother Lee Levin Werschulz
ת.ו.ש.ל.ב.ע.

—AGW

Contents

Preface

> Do not worry about your difficulties in Mathematics. I can assure you mine are still greater.
>
> Albert Einstein

Motivation and Purpose

The digital computer is one of the most transformative technologies that humanity has ever developed—perhaps, *the* most transformative. It has changed the way we learn, shop, play, interact, and think—in ways that its pioneering inventors could never have imagined. Moreover, it's a good bet that the impact computers have on our society will continue to grow in ways that we cannot even begin to imagine.

As a student at a college or university, your education should provide you with the tools to understand society—specifically, where we've been, how we got to where we are, and where we're going. It stands to reason that you should know something about computers, beyond the basics of how to use them for everyday tasks (word processing, communication, entertainment, and the like). This means that you need a somewhat deeper understanding of computers, if only to understand their inherent power, as well as their underlying limitations. In addition, just as it is important for students to understand humanities, social sciences, and fine arts, you should understand computational reasoning, which (as you will come to understand) is one of the intellectual underpinnings of our society.

Although computational reasoning is important from a strictly intellectual viewpoint, it has added importance, since practical questions that involve such reasoning arise more often than you might imagine. Here are just a few such questions:

- Does the risk associated with offshore drilling offset the benefit of obtaining additional energy from domestic sources?
- A politician is making an argument in favor of a tax cut. Is his argument valid?
- The Board of Health wants to outlaw foods that contain trans-fats. Is this policy really necessary?
- You receive an email from your best friend, saying that visiting the website `www.get-absurdly-rich-quick.com` will provide you with untold wealth, simply for the taking. Should you trust this email? Should you trust the website?
- The State Lotto jackpot is $123,000,000. Should you buy a ticket?
- You have a lot of errands to run. In what order should you make your stops, if you want to minimize your travel time?

So what should you study if you want to understand the underpinnings of computational reasoning? Not surprisingly, you should study a branch of mathematics. But what kind of mathematics will that be?

Painting with a broad brush, we see that mathematics can be divided into two streams, continuous mathematics and discrete mathematics. Most of the mathematics that you are familiar with (for instance, algebra, geometry, trigonometry, or maybe even calculus) lies within the realm of continuous mathematics, which deals with systems that can vary in an uninterrupted manner. On the other hand, you might not be very familiar with discrete mathematics, which looks very different from most of the mathematics to which you have been exposed.[1] Discrete mathematics deals with systems whose variations are disconnected from each other.

Let us try to understand how continuous and discrete mathematics differ from each other. Many of the systems found in nature are inherently continuous. For that matter, many man-made systems are also continuous; for example, if you have a dimmer attached to a light switch, if you turn down the dimmer a little, then the room will only get a little darker. But computers are different. A digital computer contains a finite amount of memory and disk space, measured in *bits*[2] ("bit" is an abbreviation for

[1] This fact may make some of you deliriously happy.

[2] Perhaps you are more familiar with the term "byte" being used to measure memory or disk capacity, as in "a 500 gigabyte hard drive." A byte is simply eight bits. If you want to know why this term came into being, go look it up some place; as of this writing, Wikipedia® had a good article about same.

"*b*inary dig*it*," meaning that a bit can represent either a zero or a one), and so a digital computer can have only a finite number of configurations (or *states*). Since a computer is a finite-state device, it is inherently discrete.

Although computers are finite-state devices, the number of states is breathtakingly large. To give you an example: I have a terabyte hard drive on my desk that I use as a backup device; a terabyte is a trillion bytes; since a byte is eight bits, this is 8 trillion bits. Depending on which way these 8 trillion bits are set to zero or one, this hard drive can be in any one of $2^{8,000,000,000,000}$ states (you will understand this better after you have studied counting in Chapter 6). In terms of the more familiar powers of ten, this number is greater than $10^{2,400,000,000,000}$. This dwarfs the number of electrons in the universe, estimated (in 1932) by British astronomer Sir Arthur Eddington, as being around 10^{79}.

In short, the very size of what we're trying to understand means that a simple case-by-case examination is out of the question. We need a more systematic way of thinking about such matters, which is provided by the tools of discrete mathematics.

It's somewhat unfortunate that the tools provided by continuous mathematics will have little use here. That's because (in all honesty) we have better tools (based on calculus, developed in the seventeenth century by Guttered Leibniz and Sir Isaac Newton) for doing continuous mathematics than discrete mathematics. On the other hand, the good news is that the tools of discrete mathematics are simpler to understand than those of continuous mathematics, since the latter require a fairly hefty intellectual overhead (namely, knowledge of calculus).

Intended Audience

> [O]f making many books there is no end; and much study is a weariness of the flesh.
>
> *Ecclesiastes 12:12*

There are hundreds of books on discrete mathematics. Why did we write one more? Very simply, we needed a discrete mathematics book for people who are not math majors. We felt that none of the available books were quite what we needed. Some books didn't have the selection of topics that we needed; some were written at the wrong level. This explains the book that you are now reading.

Our intended audience covers a fairly wide range, from people who hope to study computer or information science in great depth to people who hope to never study this material again in their lives. If there's any common factor among the students using this book, it is that they might not have a deep or broad background in mathematics. This means that to a large part, we have decided to forgo excessive mathematical formality, in the hope that this will improve your understanding of the material. (Perhaps we have bent over backwards in this direction; the word "theorem" only appears once in this book, as an historical reference on page 103.)

Topics Covered

So what topics are covered in this book and what are they good for?

Sets: *Sets* are collections of things. Sets are generally considered to be the fundamental building blocks of mathematics and that view is supported by this book— most of the topics covered in this book operate on sets.

Patterns: *Patterns*, and the ability to recognize them, have long been critical to human survival. In our modern information-based society, it is even more important to be able to identify, represent, and reason with patterns. For example, this ability can help you predict the future price of a stock and whether you should buy it or sell it. In this book we study patterns in the context of *sequences*, *summations*, and *mathematical induction*. By studying sequences and summations you will learn to both identify patterns and represent them mathematically, while mathematical induction will demonstrate how you can exploit sequential patterns to construct mathematically rigorous proofs.

Logic: The study of the principles for correct reasoning is known as *logic*. People make claims all of the time, including false claims, about everything from candy bars to political candidates. By subjecting a particular argument to the rigorous analysis provided by mathematical logic, we can often distinguish between valid and invalid claims.

Relations: Data are generally not viewed in splendid isolation; we generally seek connections. This notion is captured by the idea of a mathematical *relation* between sets. Relations are also useful to encode information, as is done in a relational database.

Functions: We often find a special kind of relation between two sets, in which a given quantity in the first set completely determines a particular quantity in the second set. Such relations are known as *functions*. Functions are important in their own right. Moreover, they capture the spirit of computation, since computation can be viewed as the act of producing a desired output from a given input value; that is, computation is inherently functional in nature.

Counting: It's important to know how to *count* accurately. As simple as this sounds, some sets are hard to count, because they are big and/or complicated. Amusingly enough, many of the classical problems that led to the discovery of combinatoric principles that allow efficient counting arise in gambling (card games, lotteries, and the like).

Probability: How can you compute the likelihood, or *probability*, that some important event will happen? Is buying that Powerball ticket a good bet? Probability allows us to reason and make decisions about events that are not assured.

Algorithms: Without software, computers would be nothing but nice-looking doorstops. Software is based on *algorithms*, which are step-by-step procedures for performing a computational task. But it is not enough to provide an algorithm to solve a task—we want algorithms to be efficient in terms of the amount of time or memory required.

Graphs: We can use *graphs* to show the connections between various objects. In particular, graphs are used to represent relations on sets. We can often rephrase a question about a relation as a question about the graph representing the relation. Sometimes, it's easy to answer such a question by simply looking at the graph, although sometimes we need to develop algorithms to answer the question.

Social Networking: A Unifying Theme

Starting in the latter part of the twentieth century, computer technology changed the way that we conduct business and live our lives. The introduction of email and simple shared electronic bulletin boards had a particularly transformative effect on how we communicate with one another. Like the postal and telephone systems of prior generations, these technological changes advanced modern communications. As people started spending more and more time connecting online via the global Internet, a new generation of users accustomed to using the computer for virtually everything evolved. Soon computers began to be used to enhance social communication in a more profound way. People used them to share their views on a wide variety of topics (such as work, sports, and politics), schedule parties, find dates, receive product recommendations, make business contacts, and stay in touch with friends and family. Thus began the revolution in *social networking*, which led to the emergence of new services such as Facebook®, Twitter®, YouTube®, LinkedIn®, and flickr®. These services allow us to share experiences after-the-fact or even moment-to-moment, and to selectively disseminate this information to our own network of friends and relatives. This revolution has fundamentally changed how we interact with others—although it is a matter of debate as to whether this is a positive development.

This book covers many of the topics that are necessary to represent these familiar social networks, and to implement the services that they provide. For example, we can represent our friends in these social networks as a *set* of people, which may be comprised of overlapping *subsets* of family, coworkers, close friends, etc. The social network will also include many *patterns* that can be exploited, if detected. We can use *logic* to make inferences and decisions about people who may not already be in our social network, to determine if they worked with us in a prior job, would make a good spouse, or have similar movie preferences. People can connect, or relate, to one another in many ways and these connections can be represented as mathematical *relations*— the most important relation for any social network site being the "friend" relation. Our friends also have an active, and changing, list of characteristics (e.g., the college from which they graduated, their relationship status, their current place of employment, and how much time they spent last week playing Farmville®). These characteristics can be viewed as *functions* since they are special relations for which there is only a single "output" value for each "input" value (e.g., every person has only one relationship status).

The ability to *count* and understand the properties of counting is also relevant to social networks. For example, LinkedIn relies on the fact that the count of the number of "colleagues of colleagues" grows very rapidly. This rapid growth increases the likelihood of a user connecting to a future employer or useful business contact via a relatively short chain of colleagues. However social networks, like life, are not all black and white, hence not all reasoning and inferences can be absolute. As such social networks must deal with *probability* and be able to reason using probabilities.

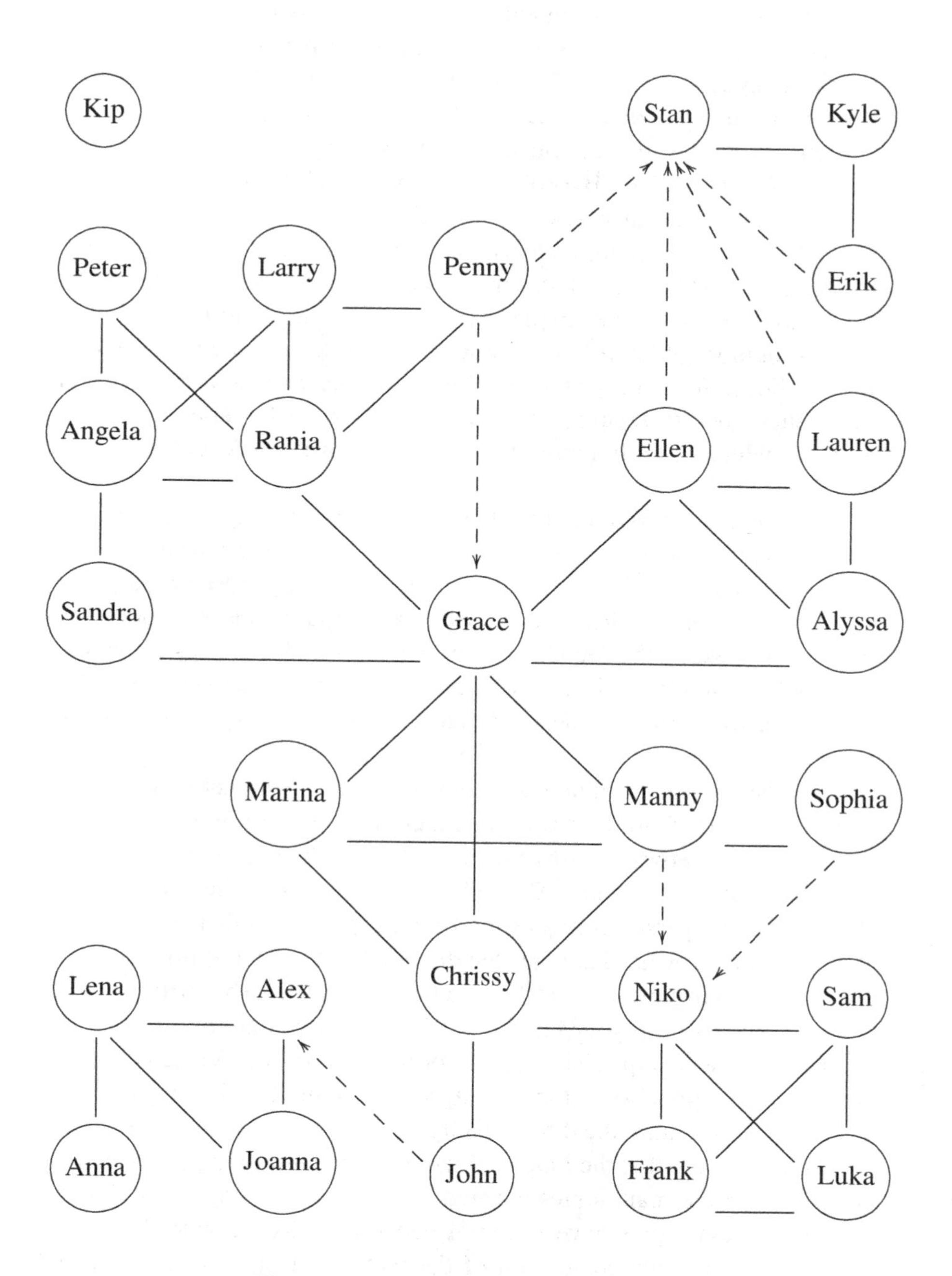

Figure 1: A sample social network

To illustrate, based on information provided, Facebook may suggest that you add Mary Smith to your network because it has reasoned that she is probably an acquaintance of yours, while eHarmony® may suggest that you arrange a date with her because she is probably a good match for you. Likewise, you may also conclude on your own that if most of your friends like the *New York Times* Bats blog, then you probably will too.

By itself, a social network is static and not very interesting. It is through the use of computer *algorithms*, which are step-by-step procedures to accomplish a task, that enable the social network to come to life and fulfill its usefulness. For example, LinkedIn uses an algorithm to find the shortest number of "hops" in your social network to another person (who may be hiring for a job you want). Finally, while a social network can be represented by sets, relations, and functions, it is ideally represented using a *graph*, which visually represents relations between objects (in this case people).

A graphical representation of a social network, similar to one you might find in Facebook, is provided in Figure 1. For the sake of convenience, we have also included this graph in the Appendix on page 315, along with several tables that describe the people in the network. This network is used as the foundation for examples and exercises throughout this book. Examples and exercises based on this social network, as well as other social networks, are denoted in this text by the **[SN]** tag, so that they are easy to identify.

As you can see in Figure 1, each user in our social network is represented as a "node" in the graph (i.e., a circle with the user's name inside it). Solid lines between pairs of nodes in the graph are used to denote that the two people are friends. Dashed lines with an arrowhead are used to denote that one person is requesting friendship with another. Note that, in this case, the arrowhead denotes the direction of the friendship request. In social networking sites, friendship is a two-way relationship without directionality (i.e., if I am your Facebook friend, then you are my friend), and as such the friendship relation does not need an arrow. Whereas a friend request is directional (i.e., I am asking you to be my Facebook friend) and therefore does require the directionality of an arrow.

You may note that the italicized terms in the preceding paragraphs are the names of the main topics covered in this book, clearly signifying the relevance these topics have to social networks. Each chapter begins by briefly reviewing the connection of the material in the chapter to social networking; moreover, each chapter will contain examples and exercises drawing from the sample social network introduced in Figure 1. In making this connection, we hope to demonstrate the relevance of the topics in this book to "the real world" or to "the real, virtual world", as the case may be.

Use of This Book

This book is designed for a one-semester undergraduate course in discrete mathematics. The book is suitable for a course offered by either a mathematics or computer science department, but the material can be given an increased computer science perspective by including the various sections that focus on computation that are distributed throughout the textbook. However, while this textbook is designed so that the majority of the content can be covered in one semester, most courses will not be able to cover quite all of the material. For this reason we provide some information on how the material in this textbook can be customized. Our experiences are influenced by the use of this book at Fordham University for two courses, Structures of Computer Science (geared towards students majoring in the humanities) and Discrete Structures (geared towards students majoring in computer science and other sciences). We typically cover about 30% more material from this book in the Discrete Structures course, including more advanced material, but we do spend more time on laboratory assignments in the Structures of Computer Science course.

Table 1 provides information that can help an instructor customize the book based on the mathematical background of the students and the pace of the intended course; we assume that those students with greater exposure to mathematics will not only be able to handle the more advanced material, but will also be able cover the material at a faster pace. The sections in Table 1 designated as "advanced" should primarily be reserved for those classes with more advanced students (the other sections can be covered by all classes). Please note, however, that in this context the term "advanced" does not imply "advanced mathematics"; rather, the term is used relative to the students with limited mathematical background who do not intend on pursuing a discipline that makes heavy use of mathematics. Also, the advanced topics can be skipped without unduly impacting the ability to understand the other sections in the book. Chapter 1, which covers sets, is the most basic chapter in the book and advanced students should be able to cover it very quickly. Table 1 also highlights those sections that especially emphasize the notion of computation. Instructors who would like to emphasize the connection between discrete mathematics and computer science should be sure to include some of these sections, while other instructors may choose to skip most or all of them. But if many of advanced sections or sections that focus on computation are covered, then it may be necessary to skip or skim some of the other material. As an example of how to use this table, an instructor that only wants to focus on the basic material and material that does not emphasize computation, should only cover the sections that do not have checkmarks in either column.

Ch	Sections	Advanced	Computation
1	Sets (all)		
2	2.1 – 2.3 Sequences and Summations		
	2.4 Mathematical Induction	✓	
3	3.1 Propositional Logic		
	3.2 Predicate Logic	✓	
4	4.1 – 4.2 Relations(basics)		
	4.3 Relational Databases	✓	✓
5	5.1–5.7 Functions (basics)		
	5.8 Password Storage Appl.	✓	✓
6	Counting (all)		
7	7.1–7.4 Probability (basics)		
	7.5 Probability Distributions	✓	
	7.6 Expected Value		
8	8.1–8.3 Algorithms (basics)		✓
	8.4 Analysis of Algorithms	✓	✓
9	9.1–9.3 Graphs (basics)		
	9.4 Minimum Spanning Tree	✓	✓
	9.5 Matrix Notation for Graphs	✓	✓

Table 1: Useful information for customizing the book

Table 1 focuses on material at the chapter or section level. Throughout the book there are a few subsections that cover supplemental material that can easily be skipped without unduly affecting he student's ability to understand the material covered in the other sections. These supplemental subsections are denoted using an asterisk ($*$). Some of the more difficult homework exercises are also denoted with an asterisk.

Support

We maintain a website

http://www.discretestructures.com

for this book. Check this site for ancillary materials, such as errata, overhead slides, and the like. Moreover, if you have any questions, compliments, or complaints about the text (of course, this includes errata), please send them to info@discretestructures.com, so that we may act on them accordingly.

Acknowledgments

The first edition of this book, as well was preliminary drafts of the first edition, were used by the Department of Computer and Information Sciences at Fordham University for its CISC 1100 (Structures of Computer Science) and CISC 1400 (Discrete Structures) courses. We are happy to thank the following current or former Fordham faculty members who have used this text in their courses or have provided suggestions about the book itself: Ying Andrews, Anthony Ferrante, Elena Filatova, Valery Frants, Eulalie Guerzon, Frank Hsu, Kevin Kelly, Yanjun Li, Chris Mesterharm, Robert Moniot, Jacek Ossowski, Christina Schweikert, Marlon Seaton, Tadeusz Strzemecki, Hoa Tran, Roger Tsai, David Wei, Larry Wolk, Xiaolan Zhang, and Yijun Zhao. In addition, we wish to thank the hundreds of students who have served as our "guinea pigs," as we refined the material in this book until it converged to its present form. Finally, we are happy to thank Patricia P. Werschulz, Esq., for legal services rendered.

Chapter 1

Sets

> My mind seems to have become a kind of machine for grinding general laws out of large collections of facts.
>
> ---
>
> *The Autobiography of Charles Darwin*

The field of discrete mathematics plays an extremely important role in the study of computer science and has contributed significantly to the advancement of modern computers and the resulting technology. Sets are fundamental to all of mathematics, therefore it is fitting that we begin our exploration into the world of discrete structures with this topic.

One of the most natural occurrences in our lives is that of a collection. If we examine our lives we will notice that we are surrounded by multiple collections. Our friends can be viewed as a collection of people, our family as another collection of people, our MP3 player as containing a collection of our favorite music, our hobbies as another collection, etc. In fact, we create collections all the time—we just don't realize that we are doing so. Thinking about which movie to see on a Friday night, we are creating a collection of potential movies from which to choose from. Thinking about what to buy our mother on Mother's Day, we encounter yet another collection, this time a collection of potential gifts that she might like. If we were to cross-reference the collection of presents we think she would like with what we can afford, we would actually be performing some kind of operation on this collection. A *set* is simply the mathematical representation of a collection and *set theory* gives us a concise and precise way to represent collections and provides a means to answer questions about them.

Let's look at our social networking example. This site contains numerous collections: people, cities, genders, birth dates, schools, and so forth. Each of these collections may be represented as a set. For instance, if we have a collection of friends from high school and a collection of friends from college, it might be important to identify those friends who overlap. Of course we know who they are, but our social networking site is also interested in who they are. Any persons in the *intersection* of your set of high school friends and set of college friends may be very important and therefore, the site could automatically flag those friends as being special.

To understand the significance that sets have within social networking, let us examine how we would build our own page on such a site. One of the most critical pieces of information that we provide our social networking site is our e-mail address. Why? Through our e-mail address, our social networking site can gain access to our set of e-mail contacts, the collection of people we communicate with. Once our e-mail contacts have been accessed, our social networking site will review our contacts, determine those who are already users in the network and will then provide us with a list of potential friends. As each of these potential friends will also have friends, these "friends of friends" can also be added to our list of potential friends, thus creating an ever expanding list of potential friends for us to request friendship with. Of course we do not have to rely solely on the set of potential friends that our social networking site provides from our contacts; we can create our own collection of potential friends. Furthermore, as we build our personal profile, we provide our site with additional information about ourselves, such as where we went to high school or college, as well as our current or previous employers. Our social networking site can then use this data to generate another list of potential friends, this time based on our own personal profile. Note that irrespective of how our list of potential friends is generated, the list stems from multiple collections of people. It is from this expansive collection that we can then choose with whom we want to request friendship. Assuming some, if not all of our requests are accepted, not only will we avoid suffering Kip's excruciating pain of having no friends,[1] but we can proudly display one large set of friends or several smaller sets each representing a specific relationship, such as family, classmates, co-workers, etc. Thus we enter the world of social networking and eventually have more friends than we thought possible. To think, all of this began from a simple collection—a *set* of people we communicate with!

[1] South Park (Season 14, Episode 4) You have 0 friends

1.1 Basic Definitions

A set is simply *a collection of objects*. The objects that comprise the set are referred to as the *members* or the *elements* of the set. For example, the set:

$$\{a, e, i, o, u\}$$

contains five elements, which (in this case) happen to be the lower-case vowels of the English alphabet.

A set must be defined precisely. Using standard mathematical notation as shown above, a set is represented using left and right curly braces ("{" and "}"). The elements of the set are placed within these curly braces and are separated by commas.

We refer to an element in a set using the membership operator, $\in$. Thus we can say that $a \in \{a, e, i, o, u\}$ since the element a is a member of the set of vowels, and we can also say that $c \notin \{a, e, i, o, u\}$ since the element c is not in the set.

A set having a finite number of elements (0, 1, 2, and so forth) is said to be finite. A set that is not finite (i.e., one with infinitely many elements) is said to be infinite. A set that has no elements is referred to as the *empty set* (or the *null set*) and is represented by $\emptyset$.

Sets represent unordered collections of distinct objects. As such, the following two rules must also be followed when defining a set:

1. The order in which you list the elements in a set does not matter.

2. The elements within a set should not be duplicated.

Thus the set $\{a, e, a, i, o, u\}$ is an improper representation of the vowel set $\{a, e, i, o, u\}$, because the element a is repeated.

Example 1.1: Is $\{\{a, b, c\}, \{b, c, a\}\}$ a proper representation of a set?
Hint: The two elements of this set are themselves sets.

Solution: Since the order of the elements in a set does not matter, sets $\{a, b, c\}$ and $\{b, c, a\}$ are identical sets. As these sets are identical and elements in a set cannot be repeated, then this is not a proper representation of a set. □

Remark. There are many mathematical applications where the order of objects does matter and in these cases sets should not be used. One instance concerns games where a player's decisions depend on the order of events. For example, if the next three cards to be dealt in a blackjack game are

10, 4, and 7, then the player will generally opt to take the third card after getting the first two cards; however, if the three cards are 10, 7, and 4, then the player will generally not take the third card after getting the first two. In cases such as these where order does matter, then we are looking at a *sequence*, rather than a set (sequences are covered in Section 2.1 of this text). □

Example 1.2: Identify which of the following sets are not properly represented.

(a) $\{\{1, 2, 3\}, \{3, 2, 1\}\}$

(b) {New York, New Jersey, Massachusetts}

(c) {1, 2, 3, Boston, {1}}

(d) {1, 2, 3, Boston, 1}

(e) $\{a, b, c, \{a, b, c\}, \emptyset\}$ □

Solution: Sets $\{\{1, 2, 3\}, \{3, 2, 1\}\}$ and {1, 2, 3, Boston, 1} are not properly represented sets, as each contain repeated elements. Note the difference in the sets represented in (c) and (d). Each of these sets contains 4 elements, but the last element in (c) is itself a set containing one element (the number 1), while the last element in (d) is the number 1. As the number 1 is listed as an element twice in this set, (d) is not valid. □

The following are a few more examples of properly represented sets:

1. $\{1, 2, 3, 4, 5, \ldots\}$
2. $\{1, 2, 3, 4, 5, 6, 7, 8, 9, 10\}$
3. $\{a, b, c, d, e, \ldots, x, y, z\}$
4. $\emptyset$
5. {Larry, Curly, Moe}
6. {Bob, red, glove, 1, 8, a, b, $\{1, 2, 3\}$, $\emptyset$, $\{\emptyset\}$}

These are all valid sets and we should be able to determine what the elements are, and in some cases, what mathematical or non-mathematical concept the set describes or represents. These six sets can be described as follows:

- The first set represents all positive integers greater than zero.
- The second set contains all integers between one and ten inclusive.
- The third set contains the lower-case letters of the English alphabet.
- The fourth set is the empty set.
- The fifth set is a set containing the names "Larry," "Curly," and "Moe."
- The sixth set is a set that, unlike all of the others, contains elements that appear to represent very different things. Like the fifth set, there is no pattern that we can identify that allows us to describe the set succinctly. Some of you may have noticed that all the elements of the fifth set are the names of the Three Stooges, so we could describe that set as such. However, how could we describe this last set, given that the elements appear to follow no pattern? We in fact have no choice but to list each element individually. The first seven elements are very basic and require no explanation. The last three elements of this set are themselves sets and can be described as follows:
  - the set containing the integers one, two and three.
  - the empty set, and finally,
  - a set containing the empty set. Note that this last element is valid and only counts as a single element. Also, this last element is *not* the empty set—it just happens to be a one-element set whose only element is the empty set.

Empty Set

The empty set can be a confusing concept, so the following conceptualization may help. Think of a set as a bag, where the matching left and right curly braces (i.e., "{" and "}") represent the bag itself and each element of the set is inside this bag. The empty set then represents an empty bag. The set $\{\emptyset\}$ represents a bag with an empty bag in it. Thus, the "outer" bag is not empty, since it has a bag inside of it.

We often need to refer to the same set several times. If the set has more than a few elements, explicitly listing all the elements of the set every time we refer to it will get pretty tiresome. For the sake of convenience, we often name our sets, generally using capital letters such as

$$A = \{a, b, c, d, e, \ldots, x, y, z\},$$
$$B = \{1, 2, 3, 4, 5, 6, 7, 8, 9, 10\}.$$

This defines two sets A and B. Set A contains all the lower-case letters of the English alphabet and set B contains the integers 1 through 10. Once the sets have been defined, they can be referred to simply as "the set A" or "the set B."

For all the sets we will be interested in, there are limits to the types of elements that can be placed in the set. For example, in some contexts we will only be interested in integer sets, while in other situations we may be interested in the set of all possible colors or all possible automobiles. The set of all possible elements that can be placed into a set is itself a set and is referred to as the *universal set*, which is generally represented by U. As we shall see, the universal set is important because certain operations, such as the *complement* of a set, can only be determined if the universal set is known.

1.2 Naming and Describing Sets

It is important to be able to describe sets formally, just as it will be important to describe sequences formally when we study them in Chapter 2. By using formal notation to describe sets, we avoid listing all the elements of a set, which would be especially time-consuming and inefficient if the sets were large (or infinite)! However, before we discuss how to describe sets, let us first introduce certain specific sets that are so common they have been assigned their own descriptive names and abbreviations. You may be familiar with many of these sets already, albeit possibly not aware they were sets. The list of most commonly used sets in mathematics are listed below:

- $\mathbb{N}$: the set of *natural numbers*. In this text,[2] the natural numbers are defined to be the set $\{0, 1, 2, 3, 4, 5, \ldots\}$ of non-negative integers.
- $\mathbb{Z}$: the set of all *integers*,[3] both positive and negative, represented by the set $\{\cdots - 3, -2, -1, 0, 1, 2, \ldots\}$.

[2] We use this definition because it is commonly used in set theory, logic, and in computer science, but sometimes the natural numbers are defined as the set of (strictly) positive integers, excluding 0 (this definition is common in number theory).

[3] Why $\mathbb{Z}$? Because the German word for "numbers" is "Zahlen."

- $\mathbb{Q}$: the set of all *rational numbers* (i.e., all numbers that can be written as *quotients*,[4] that is, fractions of the form m/n, where m and n are both integers, with $n \neq 0$. So $\frac{1}{2}$, $-\frac{1}{2}$, $\frac{2}{5}$, $\frac{12398}{45876}$, and $\frac{3}{1}$ are all elements of $\mathbb{Q}$.

- $\mathbb{R}$: the set of all *real numbers*, which may be thought of as all the points on a "number line."

The letters $\mathbb{N}$, $\mathbb{Z}$, $\mathbb{Q}$, and $\mathbb{R}$ are the standard symbols for referring to these sets, so, along with U for the universal set, these are the symbols that you should remember. There are some variations on these sets, due to additional constraints that we can add. These are shown below via some examples. This notation is not necessarily standard, but is used because it is quite clear.

- $\mathbb{Z}^+$ is the set of positive integers.

- $\mathbb{Z}^-$ is the set of negative integers.

- $\mathbb{Z}^{\geq 0}$ is the same as $\mathbb{N}$.

- $\mathbb{R}^{>7}$ is the set of all real numbers greater than seven.

Remark. The definition of $\mathbb{R}$ that we gave may strike you as being somewhat imprecise, as compared that that of (say) $\mathbb{Q}$. That's because the notion of "real number" is fairly sophisticated. It wasn't until 1871 that a precise definition of $\mathbb{R}$ was given. Interestingly enough, it was given by by Georg Cantor, the inventor of set theory. □

In some of the previous examples we referred to a set by referring to its elements and, in some cases had to list them individually. Clearly this is impractical for large sets and impossible for sets with an infinite number of elements. *Set builder notation* is a specific language construct that was developed to provide an easier and more concise method for defining new sets. This is best understood with an example. Let us say we wanted to define a set A that contains all positive even integers. To do so we could do one of the following:

- Specifically state that set A is comprised of all positive even integers.

- Explicitly list the positive even integers, as in

$$A = \{2, 4, 6, 8, 10, 12, 14, 16, 18, \dots\}.$$

[4] The $\mathbb{Q}$ should remind you of the word "quotient."

- Use set builder notation to specify set A.

We can use set builder notation to define the set A of even numbers as follows (this list is not exhaustive):

$$A = \{x : x \text{ is a positive even integer}\}$$
$$A = \{x \mid x \text{ is a positive even integer}\}$$
$$A = \{x : x \in \mathbb{Z} \text{ and } x \text{ is positive and } x \text{ is even}\}$$
$$A = \{x \mid x \in \mathbb{Z} \text{ and } x \text{ is positive and } x \text{ is even}\}$$
$$A = \{x \in \mathbb{Z} : x \text{ is a positive even integer}\}$$
$$A = \{x \in \mathbb{Z} \mid x \text{ is a positive even integer}\}$$
$$A = \{x \in \mathbb{Z}^+ : \ x \text{ is even}\}$$
$$A = \{x \in \mathbb{Z}^+ \mid \ x \text{ is even}\}$$
$$A = \{x : x = 2y \text{ for } y \in \mathbb{Z}^+\}$$
$$A = \{x \mid x = 2y \text{ for } y \in \mathbb{Z}^+\}$$

This may look complicated at first, but once we understand how to translate this notation, it will be clear that each of these expressions represents the set of all positive even integers.

To understand this translation we can think of set builder notation as an expression comprised of two parts typically separated by either the ":" or "|" symbol. The first part of the set builder notation describes the elements to be included in the set, whereas the second part places specific constraints on those elements, determining the actual values that will be included. The symbol ":" or "|" that we use to separate these two parts represents the phrase "such that." With this in mind, the set builder notation that describes our example can be read as

"all elements x *such that*
x is an element of the positive integers and x is even"

or as

"all elements x in the set of positive integers *such that*
x is even."

For our example:

- The first part names a variable x, which acts as a place holder representing the values that may be attained by elements of our set. We often tighten things up a bit by further requiring these values to belong to a known, pre-existing set; for this example, this set was either the integers $\mathbb{Z}$ or the positive integers $\mathbb{Z}^+$.

- In the second part, each of the following expressions can be used because they each enforce the constraint that the variable x must be an even positive integer:
  - x is a positive even integer,
  - $x \in \mathbb{Z}$ and x is positive and x is even,
  - $x \in \mathbb{Z}^+$ and x is even, and
  - $y \in \mathbb{Z}^+$ and $x = 2y$.

Remark. In the example given above, we always used x to name the variable appearing in the left-hand side of the set builder notation describing our set. There's nothing special about the letter x; we could've used any other letter as well, provided that we were consistent in our usage. So we could also have (for example):

$$A = \{\, z : z \text{ is a positive even integer} \,\}$$
$$A = \{\, m : m \in \mathbb{Z} \text{ and } m \text{ is positive and } m \text{ is even} \,\}$$
$$A = \{\, q \in \mathbb{Z}^+ \mid q \text{ is even} \,\}$$
$$A = \{\, s : s = 2t \text{ for } t \in \mathbb{Z}^+ \,\}.$$

□

Example 1.3: Express the set $A = \{1, 2, 3, 4, 5\}$ using set-builder notation.

Solution: The following is just one way to describe the set using set builder notation. Can you think of other ways?

$$A = \{x \mid x \in \mathbb{Z} \text{ and } 1 \leq x \leq 5\}.$$

□

Example 1.4: Give an explicit listing for each of the following sets, which are described via set builder notation.

(a) $A = \{\, x \mid x \in \mathbb{Z} \text{ and } 10 \leq x \leq 100 \,\}$.

(b) $B = \{\, y : y = k + 1 \text{ and } k \in \{5, 6, 7\} \,\}$.

(c) $C = \{\, z \mid z \in \mathbb{Z} \text{ and } z \leq 1000 \,\}$.

(d) $D = \{\, s : s \in$ "U.S. states beginning with the letter B" $\}$

(e) $E = \{\, c \mid c \in$ "all courses in your current semester course schedule" $\}$.

(f) $F = \{\, m \in \mathbb{Z}^- \mid m \text{ is odd} \,\}$

Solution:

(a) $A = \{10, 11, 12, 13, \ldots, 100\}$

(b) $B = \{6, 7, 8\}$

(c) $C = \{\ldots, -3, -2, -1, 0, 1, \ldots, 999, 1000\}$

(d) $D = \emptyset$

(e) This might be

$$E = \{\text{computer science, physics, english, history, philosophy}\},$$

as just one example.

(f) $F = \{\ldots - 9, -7, -5, -3, -1\}$ □

1.3 Comparison Relations on Sets

We can compare sets to one another. For example, we can ask if one set is equal to another or if one set contains another. This is similar to asking if one number equals another or if one number is less than another. This is accomplished using special relational operators (we discuss relations in depth in Chapter 4). In this section we discuss the relational operators for sets.

The most basic relational operator is equality. We say two sets A and B are *equal* (i.e. $A = B$) if they have exactly the same elements. If two sets A and B are not equal, we write $A \neq B$. Notice that we use the same notation for set equality (and inequality) as we do with numbers.

Fact. *For any set A, the statement $A = A$ is always true, and the statement $A \neq A$ is always false.* □

Continuing with the idea that set comparisons are analogous to comparisons between numbers, we now define two additional ways of comparing sets, which will be analogous to the $\leq$ and $<$ numerical relations.

Suppose that A and B are sets, with every element of A also being an element of B. Then A is said to be a *subset* of B, which we write as $A \subseteq B$. When $A \subseteq B$, we sometimes say that A is *included* in B. As an example, let $T = \{a, b, c, \ldots, x, y, z\}$. Then we can say:

$$\{a, b\} \subseteq T,$$
$$\{a, e, i, o, u\} \subseteq T,$$
$$\{l, k, r\} \subseteq T.$$

It will not surprise you too much to learn that if $A \subseteq B$ is not true (i.e., if some element of A is not an element of B), then we write $A \nsubseteq B$. Again letting $T = \{a, b, c, \ldots, x, y, z\}$, we see that

$$\{a, b, c, 5\} \nsubseteq T,$$
$$\{10, 9, 8\} \nsubseteq T,$$
$$\{\{a, b, c, \ldots, x, y, z\}\} \nsubseteq T.$$

Take a closer look at the set $\{\{a, b, c, \ldots, x, y, z\}\}$. Notice that this is a one-element set, its sole element being a set itself, namely, the set of the lower-case letters of the English alphabet. This is different from our set T, which is a set containing all the lower-case letters of the English alphabet. This distinction is further highlighted by noting that while the set T contains 26 elements, the set $\{\{a, b, c, \ldots, x, y, z\}\}$ contains only 1 element.

Just as "$<$" is a special instance of "$\leq$," there's a corresponding special case "$\subset$" of "$\subseteq$." Suppose that A and B are two sets, with $A \subseteq B$ but $A \neq B$; that is every element of A is an element of B, but there is at least one element of B that is not an element of A. Then we say that A is a *proper subset* of B, denoted $A \subset B$. Of course, if $A \subset B$ does not hold, we will write $A \not\subset B$.

Remark. The relationship between $\subset$ and $\subseteq$ (for sets) is similar to that between $<$ and $\leq$. Just as the line beneath the "$<$" part of the "$\leq$" symbol reminds us that the two numbers may be equal, the line below the "$\subset$" part of the "$\subseteq$" symbol indicates that the sets may be equal. □

Fact. *For any sets S and T, the following statements are always true:*

1. $S = T$ *if and only if* $S \subseteq T$ *and* $T \subseteq S$.

2. $S \subseteq S$.

3. $S \not\subset S$.

4. $\emptyset \subseteq S$. □

Why are these statements true?

- Statement 1 is true because if we have two sets and every element in one set is a member of the other, and vice versa, then both sets must contain the same elements.

- Statement 2 is true because every element of S on the left-hand side of the statement must be an element of S (on the right-hand side).

- Statement 3, which says that S is not a proper subset of itself, is true because while a set can be a subset of itself, it cannot be a proper subset of itself (to be a proper subset the set on the right-hand side must have at least one element that is not in the set on the left-hand side, but this cannot happen since the sets are equal).

- Statement 4 is true but this may not be obvious. Here's an explanation that works for finite sets (for the general case, see the Remark on page 83). Take the set S and start removing elements, one by one. No matter how many elements you remove, the smaller set is still a subset of S, because all of its elements must be in S (taking away elements cannot make this untrue). After you've removed all the elements, you're left with the empty set, which therefore must be a subset of S.

Example 1.5: Let $B = \{a, b, c, d\}$. Which of the following statements are true?

(a) $\{a, b\} \subseteq B$

(b) $\{a, b\} \subset B$

(c) $a \subseteq B$

(d) $\{a, b, c, d\} \subset B$

(e) $B \subseteq B$

(f) $\emptyset \subset B$

(g) $B \subset \emptyset$

Solution: Statements (a) and (b) are both true. Statement (c) is not true since only a set can be a subset and a is not a set. Statement (d) is not true since $\{a, b, c, d\}$ is a subset but not a proper subset. From the fact given above that deals with subset inclusion and the like, we know that statements (e) and (f) are true. Statement (g) is false since not every element in set B is in $\emptyset$ (in fact none of them are in the empty set). □

Example 1.6: Let

$$A = \{\text{purple, blue, fuchsia, red}\} \qquad \text{and} \qquad B = \{\text{blue}\}.$$

Fill in the missing symbol(s) from the set $\{\in, \subseteq, \subset, =\}$ to correctly complete each of the following statements:

(a) B ___ A

(b) blue ___ A

(c) green ___ A

(d) A ___ A

(e) A ___ B

Solution: We represent each answer as a set, which represents the set of all symbols that may be inserted correctly into the blank (we told you that sets would be useful). The answers are listed in order.

(a) $\{\subset, \subseteq\}$

(b) $\{\in\}$

(c) None of these. The most meaningful thing we could say is that green $\notin A$.

(d) $\{\subseteq, =\}$

(e) None of these. However, it *is* meaningful to say that $A \nsubseteq B$, from which it follows that $A \not\subset B$ and that $A \neq B$. □

Example 1.7: **[SN]** Using our social network diagram (on page 317), do Grace, Lauren, Alyssa, Niko, Michael represent a subset of users of our social network?

Solution: Let

$$U = \{\text{all the users of our social network}\}$$
$$A = \{\text{Grace, Lauren, Alyssa, Niko, Michael}\}$$

Although the set {Grace, Lauren, Alyssa, Niko} $\subset U$, we have $A \not\subset U$, since Michael is not a user in our social network. □

1.4 Set Operators

Just as we can perform mathematical operations like, addition, subtraction, multiplication, and division on known numbers to get new numbers, we can also get new sets by performing set operations on known sets. These set operations are called *union*, *intersection*, *difference*, *complement*, *power set*, and *Cartesian product*.

Before examining these operations, we will first introduce the idea of a set's *cardinality*, which is simply the number of elements that the set contains. The cardinality of the set S is denoted by $|S|$. For example:

- If $A = \{2, 5, 8, 11, 14\}$, then $|A| = 5$.
- If $B = \{\text{circle, rectangle, octagon}\}$ then $|B| = 3$.

The reason for this notation is that $|S|$ is the size of the set S, just as the absolute value $|a|$ is the size of the number a.

Once the cardinality of a set is determined, mathematical operations such as addition and subtraction can be performed on the cardinality of those sets, as shown in the following example:

Example 1.8: Suppose that

$$\begin{aligned} A &= \{\text{gold, silver, bronze}\} \\ B &= \{-3, 0, 3, 10\} \\ C &= \{\{x, \{y\}\}, \{x, y, z\}\} \\ D &= \{\emptyset, \text{circle, square}\} \end{aligned}$$

Determine the value of $|A|$, $|B| + |C|$, and $|A| + |B| - |D|$.

Solution: We have

$$\begin{aligned} |A| &= 3 \\ |B| + |C| &= 4 + 2 = 6 \\ |A| + |B| - |D| &= 3 + 4 - 3 = 4 \end{aligned}$$

□

Example 1.9: [SN] Using our social network diagram (on page 317), how many friends are requesting friendship with Stan?

Solution: We can see easily see from the diagram that Erik, Lauren, Ellen and Penny are each requesting friendship with Stan. We can represent this as $F = \{\text{Erik, Lauren, Ellen, Penny}\}$ and then $|F| = 4$, which provides us with the answer we are looking for. □

Example 1.10: [SN] How can we use the cardinality of a set to determine if each of our users in our social network diagram (on page 317), is currently living in their hometown?

Solution: Using the information provided in Table A.1 (on page 318), create a set for each one of our users that contains as elements their hometown and the city they are currently living in. Example:

$$G = \{\text{Hartford, Boston}\}$$
$$A = \{\text{Hartford}\}$$

are the sets that represent this information for our users Grace and Anna respectively. We can see from these sets that $|G| = 2$ and $|A| = 1$. Given that elements in a set cannot be duplicated (see Section 1.1), we can deduce solely from the cardinality of these sets that Anna currently lives in her hometown whereas Grace does not. □

Venn diagrams are a useful tool for studying set operations as they provide a clear visual representation of the ensuing sets. In a Venn diagram, the universal set is represented by a rectangular region and the subsets of the universal set are typically represented as interlocking circles drawn within the rectangular region. A typical Venn diagram is shown in Figure 1.1. With this in mind let's continue with the set operations *union*, *intersection*, *difference*, *complement*, *power set*, and *Cartesian product*.

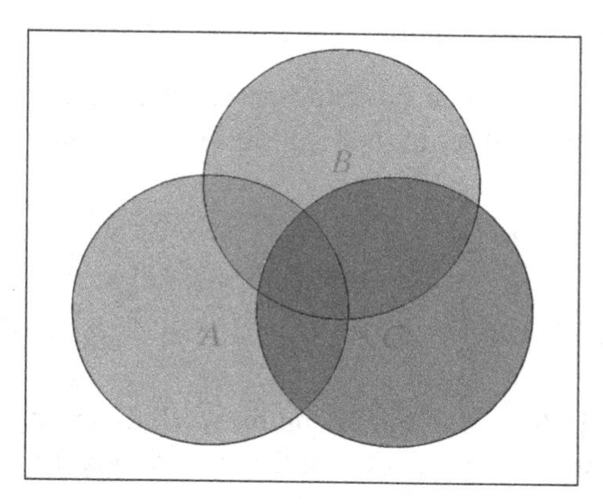

Figure 1.1: A typical Venn diagram

The *union* of two sets A and B, denoted $A \cup B$, is a new set whose elements are all the elements in set A together with all the elements in set B, so that

$$A \cup B = \{x : x \in A \text{ or } x \in B\}.$$

Note that as elements in a set cannot be repeated, if the sets forming the union contain common elements, those elements should only be included once. The Venn diagram in Figure 1.2 illustrates the union of two sets.

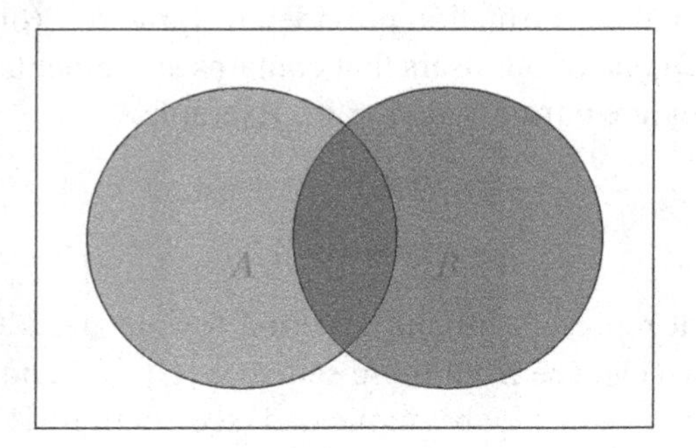

Figure 1.2: The entire shaded area represents the union $A \cup B$

Example 1.11: Let

$$\begin{aligned} A &= \{0, 1, 2, 3, 4, 5, 6\}, \\ B &= \{2, 4, 6, 8, 10\}, \\ C &= \{0, 5, 10, 15\}, \\ D &= \emptyset. \end{aligned}$$

What are $A \cup B$, $C \cup D$, and $(A \cup B) \cup (C \cup D)$?

Solution: The answers can be determined by combining the elements in the appropriate sets and removing any duplicates. Thus, we get

$$\begin{aligned} A \cup B &= \{0, 1, 2, 3, 4, 5, 6, 8, 10\} \\ C \cup D &= \{0, 5, 10, 15\} \\ (A \cup B) \cup (C \cup D) &= \{0, 1, 2, 3, 4, 5, 6, 8, 10, 15\} \end{aligned}$$

□

Example 1.12: [SN] Using our social network diagram (on page 317), let M, G and N represent the set of (confirmed) friends for Manny, Grace and Niko respectively. What is $M \cup G \cup N$?

Solution: Create the set of friends for Manny, Grace and Niko as:

M = {Grace, Marina, Chrissy, Sophia}
G = {Ellen, Rania, Sandra, Alyssa, Marina, Chrissy, Manny}
N = {Chrissy, Sam, Frank, Luka}

Form the union between two of the three sets, say $M \cup G$, then form the union between this resulting set and the third set, N in this case:

$$M \cup G = \{\text{Grace, Marina, Chrissy, Sophia, Ellen, Rania, Sandra, Alyssa, Manny}\}$$

$$M \cup G \cup N = \{\text{Grace, Marina, Chrissy, Sophia, Ellen, Rania, Sandra, Alyssa, Manny, Sam, Frank, Luka}\}$$ □

The *intersection* of two sets A and B, denoted $A \cap B$, is a new set whose elements are all the elements that the sets A and B have in common, so that

$$A \cap B = \{x : x \in A \text{ and } x \in B\}.$$

The Venn diagram in Figure 1.3 illustrates the intersection of two sets.

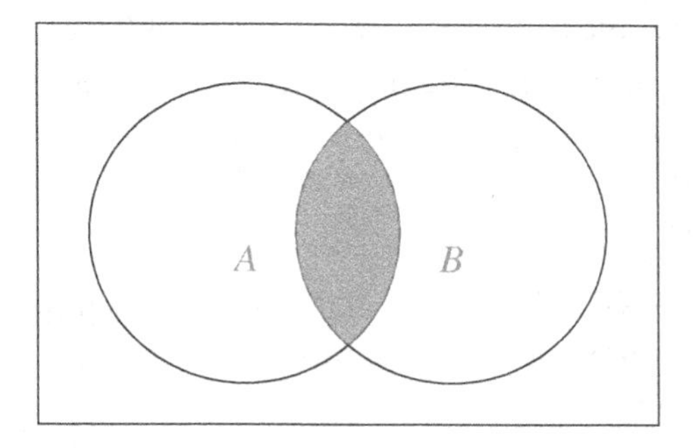

Figure 1.3: The shaded region in the center represents $A \cap B$

Example 1.13: Let the sets A, B, C, and D be defined as in Example 1.11.

$$A = \{0, 1, 2, 3, 4, 5, 6\},$$
$$B = \{2, 4, 6, 8, 10\},$$
$$C = \{0, 5, 10, 15\},$$
$$D = \emptyset,$$

What are $A \cap B$, $C \cap B$, $(A \cap B) \cup (C \cap B)$, and $(A \cup B) \cap D$?

Solution: We easily find that

$$A \cap B = \{2, 4, 6\},$$
$$C \cap B = \{10\},$$
$$(A \cap B) \cup (C \cap B\} = \{2, 4, 6, 10\},$$
$$(A \cup B) \cap D = \emptyset.$$ □

If the intersection of two sets results in the empty set, then the sets are considered to be *disjoint*. In other words, if sets A and B are disjoint, then they have no common elements.

Example 1.14: As in Example 1.11, let

$$A = \{0, 1, 2, 3, 4, 5, 6\},$$
$$B = \{2, 4, 6, 8, 10\},$$
$$C = \{0, 5, 10, 15\}.$$

Show that $A \cap B$ and C are disjoint.

Solution: Since $A \cap B = \{2, 4, 6\}$ and $C = \{0, 5, 10, 15\}$, we see that $(A \cap B) \cap C = \emptyset$. Thus $A \cap B$ and C are disjoint. □

Example 1.15: If two sets S and T are disjoint, would the Venn diagram of $S \cap T$ look the same as if they were not disjoint?

Solution: As the sets S and T are disjoint, they share no common elements and therefore, $S \cap T$ would produce the empty set. With respect to the Venn diagram, as the intersection of these two sets is the empty set, the circles would *not* intersect. □

Example 1.16: [SN] Using our social network diagram (on page 317), let M, G and N represent the set of friends for Manny, Grace and Niko respectively. What is $|M \cap G \cap N|$?

Solution: As with Example 1.12, create the set of friends for Manny, Grace and Niko as:

$$M = \{\text{Grace, Marina, Chrissy, Sophia}\}$$
$$G = \{\text{Ellen, Rania, Sandra, Alyssa, Marina, Chrissy, Manny}\}$$
$$N = \{\text{Chrissy, Sam, Frank, Luka}\}$$

Similarly, form the intersection with two of the three sets, say M and G, and intersect this resulting set with set N as:

$$M \cap G = \{\text{Marina, Chrissy}\}$$
$$M \cap G \cap N = \{\text{Chrissy}\}$$

Since there is one element remaining in the final set $|M \cap G \cap N| = 1$. □

The *difference* of two sets A and B, denoted $A - B$, is a new set whose elements are all the elements in set A that are not in set B, so that

$$A - B = \{x : x \in A \text{ and } x \notin B\}.$$

The Venn diagram of Figure 1.4 illustrates the difference of two sets.

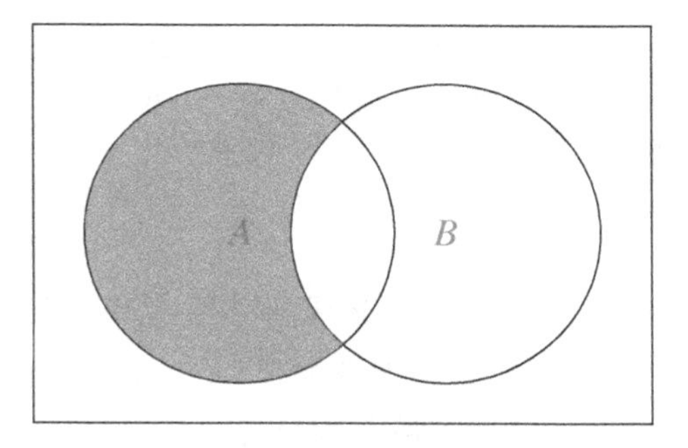

Figure 1.4: The shaded region represents $A - B$

Example 1.17: Given the sets

$$A = \{0, 1, 2, 3, 4, 5, 6\}$$
$$B = \{2, 4, 6, 8, 10\}$$
$$C = \{0, 5, 10, 15\},$$

what are $A - B$, $C - B$, $(A - B) - (C - B)$, $A - C$, and $B - A$?

Solution: We have

$$A - B = \{0, 1, 3, 5\},$$
$$C - B = \{0, 5, 15\},$$
$$(A - B) - (C - B) = \{1, 3\},$$
$$A - C = \{1, 2, 3, 4, 6\},$$
$$B - A = \{8, 10\}.$$

Note that $(A - B) - (C - B) \neq A - C$ (no "cancellation"); moreover, $A - B \neq B - A$ (set difference is not commutative). Thus the order in which the sets appear in the statement matters. □

Example 1.18: [SN] Notice in Example 1.12 (on page 16), when we formed the union of the friends of Manny, Grace and Niko, that Manny and Grace were also included in the resulting set as they themselves are friends. How could the difference operator be used to create a set representing the union of all the friends of these three users, but not contain any one of them?

Solution: Using sets M, G, and N as defined in Example 1.12 additionally let,

$$F = \{\text{Manny, Grace, Niko}\}.$$

The solution to our problem is given by $(M \cup G \cup N) - F$. □

Our next operation requires us to have a given fixed universal set U. Let A be a subset of U. The *complement* A' of A is the set of elements from the universal set U that are not in A, so that

$$A' = U - A.$$

The Venn diagram in Figure 1.5 illustrates the complement of a set.

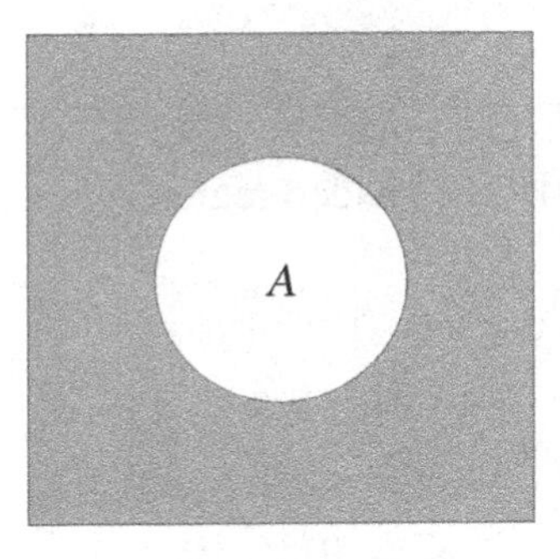

Figure 1.5: The shaded region represents A'

Example 1.19: Given the universal set

$$U = \{a, b, c, d, e, f, g, \ldots, x, y, z\},$$

find A' for the set

$$A = \{a, e, i, o, u\}.$$

Solution: The set A' is made up of the elements in U that are not in A. Thus we get

$$A' = \{b, c, d, f, g, h, j, k, l, m, n, p, q, r, s, t, v, w, x, y, z\}.$$ □

Example 1.20: Let set A be a subset of some universal set U. What do the two expressions $A \cup A'$ and $A \cap A'$ evaluate to?
Solution: We find that

$$A \cup A' = U$$
$$A \cap A' = \emptyset$$ □

Example 1.21: [SN] Using our social network diagram (on page 317) and the set of professions listed in Table A.1 (on page 318), how could we use the complement operator to represent the set of professions not represented by any of our users?

Solution: Let

$$U = \{\text{universal set of all possible professions}\}$$
$$P = \{\text{set of all professions represented by our users}\}$$

P' represents the set of all professions not represented by the users in our social network. □

Example 1.22: (This example shows how we can combine set operations.) On a typical Friday night a group of friends is deciding whether to order Chinese, Greek or Italian food. Assuming that everybody involved wants to eat (at least) one of these fine cuisines, how would you represent the following using set notation?

(a) The universal set.

(b) The number of friends in the universal set.

(c) The set of friends who are willing to order Greek and Chinese food.

(d) The set of friends who are willing to order Chinese or Italian food.

(e) The set of friends who will not order Italian food.

(f) The number of friends who are willing to order Italian or Greek food, but not Chinese food.

(g) The set of friends who will eat any of these three cuisines.

(h) The set of friends who will eat only one of these kinds of food.

(i) The number of friends who will only order Italian food.

Solution: Let C be the set of friends who are willing to order Chinese food, G be the set of friends who are willing to order Greek food, and I be the set of friends who are willing to order Italian food.

(a) $U = C \cup G \cup I$

(b) $|U|$

(c) $G \cap C$

(d) $C \cup I$

(e) I'

Name	Law
Identity	$S \cap U = S$
Identity	$S \cup \emptyset = S$
Complement	$S \cap S' = \emptyset$
Complement	$S \cup S' = U$
Double Complement	$(S')' = S$
Idempotent	$S \cap S = S$
Idempotent	$S \cup S = S$
Commutative	$A \cap B = B \cap A$
Commutative	$A \cup B = B \cup A$
Associative	$(A \cap B) \cap C = A \cap (B \cap C)$
Associative	$(A \cup B) \cup C = A \cup (B \cup C)$
Distributive	$A \cap (B \cup C) = (A \cap B) \cup (A \cap C)$
Distributive	$A \cup (B \cap C) = (A \cup B) \cap (A \cup C)$
DeMorgan	$(A \cap B)' = A' \cup B'$
DeMorgan	$(A \cup B)' = A' \cap B'$
Equality	$A = B$ if and only if $A \subseteq B$ and $B \subseteq A$
Transitive	if $A \subseteq B$ and $B \subseteq C$, then $A \subseteq C$

Table 1.1: Some handy laws of set theory

(f) $|C'|$

(g) $C \cap G \cap I$

(h) $(C - (G \cup I)) \cup (G - (C \cup I)) \cup (I - (C \cup G))$

(i) $|I - (C \cup G)|$ □

When working with numbers or algebra, you probably learned a few laws or identities that simplified your work. So, you may not be surprised to know that set theory has some identities as well. A small sampling may be found in Table 1.1. Here, U is the universal set, with $A, B, C, S \subseteq U$.

For our next operation, we need to introduce the notion of ordered pairs. Let A and B be sets. Pick two elements, $a \in A$ and $b \in B$. Then (a, b) denotes an *ordered pair*, whose first element is a, and whose second element is b.

Warning: Although order does not matter when you're specifying sets, the order does matter when you're looking at ordered pairs. For example, let $A = B = \mathbb{R}$, so that we're looking at ordered pairs of real numbers.

Although $\{3, 4\} = \{4, 3\}$, the ordered pairs $(3, 4)$ and $(4, 3)$ are not equal. □

The set of all ordered pairs, in which the first element comes from the set A and the second from the set B, is called the *Cartesian product* or *cross product* of A and B, denoted $A \times B$, so that

$$A \times B = \{(x, y) \mid x \in A \text{ and } y \in B\}.$$

As always, this concept is best understood with a few examples.

Example 1.23: Given that

$$A = \{1, 2, 3\}$$
$$B = \{a, b, c\}$$
$$C = \{-1, z\},$$

determine the Cartesian products $A \times B$, $B \times A$, $C \times A$, and $B \times C$.

Solution: We have

$$A \times B = \{(1, a), (1, b), (1, c), (2, a), (2, b), (2, c), (3, a), (3, b), (3, c)\}$$
$$B \times A = \{(a, 1), (a, 2), (a, 3), (b, 1), (b, 2), (b, 3), (c, 1), (c, 2), (c, 3)\}$$
$$C \times A = \{(-1, 1), (-1, 2), (-1, 3), (z, 1), (z, 2), (z, 3)\}$$
$$B \times C = \{(a, -1), (a, z), (b, -1), (b, z), (c, -1), (c, z)\}$$ □

Example 1.24: [SN] One evening Grace, one of our more connected users on our social network (on page 317), decided to play match maker. Based on all the profiles of friends and friends of friends she was able to access, she determined that Sandra, Rania and Angela were available females and Peter and Frank were available males. How many ways can she introduce her female "friends" to her male "friends"?

Solution: To answer this question we must pair up each female friend with each male friend, essentially forming a Cartesian product of two sets; Grace's available female friends and Grace's available male friends. Therefore, let

$$F = \{\text{Sandra, Rania, Angela}\}$$
$$M = \{\text{Peter, Frank}\}$$

then

$$F \times M = \{(\text{Sandra,Peter}), (\text{Sandra,Frank}), (\text{Rania, Peter}), (\text{Rania, Frank}), (\text{Angela, Peter}), (\text{Angela, Frank})\}$$

Since $|F \times M| = 6$, Grace has six possible ways to introduce her available female friends to her available male friends. □

Remark. Why do we use the adjective "Cartesian" to describe the set product? Let's think about $\mathbb{R} \times \mathbb{R}$, often denoted $\mathbb{R}^2$ for short.[5] From the definition, we see that

$$\mathbb{R}^2 = \{ (x, y) : x \in \mathbb{R} \text{ and } y \in \mathbb{R} \}.$$

We see that $\mathbb{R}^2 = \mathbb{R} \times \mathbb{R}$ can be identified with the set of all points in the plane, with (x, y) denoting the point whose horizontal coordinate is x and whose vertical coordinate is y. This technique for identifying points in the plane was discovered in 1637 by the French mathematician and philosopher[6] René Descartes, and is called the Cartesian coordinate system, in his honor. So the Cartesian plane $\mathbb{R}^2$ is simply the Cartesian product $\mathbb{R} \times \mathbb{R}$. □

It is easy to determine the cardinality of a set product. Given two sets X and Y, the cardinality of the Cartesian product will always equal the product of the cardinalities of the two sets. That is, for any sets A and B, we have

$$|A \times B| = |A| \cdot |B|. \tag{1.1}$$

(This relationship explains why we call $A \times B$ the "product" of the sets A and B.) It is easy to see why this formula is true; since we can pair each element in A with each element in B, there are $|A| \cdot |B|$ pairs in all. In Chapter 6, we shall see that this formula is the basis of the *multiplication rule*,which is useful for determining the size of large, complicated sets (such as the possible outcomes in various games of chance).

Example 1.25: Mother's Day is just around the corner and you need to decide on a present for mom. You realize that although she told you that "the only present I want is to spend the day with you," you're toast if you don't come up with something nice. Suppose that her wish list consists of a wallet, perfume and a new book by her favorite author. You are also considering flowers or a gift certificate. The total amount of money you have to spend is $28.00. Assuming that the cost of each item you are considering is:

- wallet: $125.00
- perfume: $79.00

[5] After all, squaring is nothing other than multiplication in which the two factors are the same.

[6] As in "Cogito ergo sum", i.e., "I think, therefore I am."

- book: \$24.00
- flowers: \$22.00
- gift certificate: \$20.00

How will you choose which present to purchase?

Solution: You may wonder about the point of this question. Clearly, you can easily determine that you can only afford to purchase either the book, flowers, or gift certificate. You may recall that in the introduction of this chapter it was stated that we form collections (i.e., sets) and perform operations on those collections all the time. This example is intended to illustrate this point.

It should be obvious that our choice of presents is actually a set, which we can represent as,

$$P = \{\text{wallet, perfume, book, flowers, gift certificate}\}$$

or more specifically as,

$$P = \{125, 79, 24, 22, 20\}$$

since your decision will be made based on the cost of each item. Less obvious is that the amount of money you have to spend is also a set, which we can represent as

$$M = \{28\}.$$

You may not realize it, but in determining which present you can afford to buy, you are creating a cross reference between the amount of money you have available and the cost of each present on your list. In essence you are performing the Cartesian product

$$M \times P = \{(28, 125), (28, 79), (28, 24), (28, 22), (28, 20)\}$$

where the first entry from each pair represents the amount of money you have available and the second entry represents the cost of the present you are considering. From this set of choices, you then select the choices that are viable (i.e. pairs (x, y) for which $x \geq y$), creating *another* set R that represents *only* those viable choices, so that

$$R = \{(28, 24), (28, 22), (28, 20)\}.$$

Your choice of presents is then one of,

$$\{\text{book, flowers, gift certificate}\}.$$

Notice that your decision is based on the relationship that exists between the amount of money you have available and, the cost of the choice of presents. Relations between sets is a topic that will be discussed in more detail in Chapter 4. □

The last operation we shall describe is somewhat subtle. Given a set A, we can create a new set whose elements consist of all possible subsets of the original set A, including the empty set $\emptyset$. This new set is referred to as the *power set* of the *base set* A. We denote the power set of A as $\mathscr{P}(A)$. Thus

$$\mathscr{P}(A) = \{ X : X \subseteq A \},$$

which means that

$$X \in \mathscr{P}(A) \qquad \text{if and only if} \qquad X \subseteq A.$$

Note that the power set is a set of sets!

It's very easy to get lost in the details of computing a power set; in particular, it's very easy to miss one of the subsets of the base set. If we are to cope with the complexity of this task, we need to be organized. Suppose that we want to compute $\mathscr{P}(X)$, where X is a set of cardinality n. To proceed systematically, we recommend that you list

- all subsets of cardinality 0 (there's only one, namely, the empty set), followed by
- all subsets of cardinality 1 (there will be one n such subsets, one for each element of X),
- all subsets of cardinality 2 (there will be $\frac{1}{2}n(n-1)$ such subsets, one for each distinct pair of elements from X—see Chapter 6 if you want to know how to figure this out),
- ... working our way up to ...
- all subsets of cardinality $|n|$ (there's only one, namely, X itself).

Let's do a few examples, to see how this works.

Example 1.26: If the set $X = \{\text{circle, square}\}$, then what is $\mathscr{P}(X)$?

Solution: Following the prescription given above, we see that we need to list the following subsets of X:

- One subset of cardinality 0: the empty set $\emptyset$.

- Two subsets of cardinality 1: {circle} and {square}.
- One subset of cardinality 2: {circle, square}.

Therefore $\mathscr{P}(X) = \{\emptyset, \{\text{circle}\}, \{\text{square}\}, \{\text{circle, square}\}\}$. □

Example 1.27: If the set $Y = \{a, b, c, d\}$, what is $\mathscr{P}(Y)$?

Solution: Listing the subsets of cardinality 0, 1, 2, 3 and 4, we find that

$$\mathscr{P}(Y) = \{\emptyset, \{a\}, \{b\}, \{c\}, \{d\}, \{a, b\}, \{a, c\}, \{a, d\}, \{b, c\}, \{b, d\}, \{c, d\}, \{a, b, c\}, \{a, b, d\}, \{a, c, d\}, \{b, c, d\}, \{a, b, c, d\}\}.$$ □

Example 1.28: [SN] Using our social network diagram (on page 317), how many possible ways can the friends Niko, Sam, Frank, Luka set up a chatting session on any given day?

Solution: You will see more of these types of problems when we study counting in Chapter 6, but for now, thinking about the various ways this can be accomplished we see that there may be no chatting session, each of the friends can be on by themselves - waiting, all four friends can be on chatting together, 2 of the 4 friends can be chatting or 3 of the 4 friends can be chatting. Although we may not realize it, essentially what we are doing is forming a power set of these four friends as:

$\mathscr{P}(\{\text{Niko, Sam,Frank, Luka}\}) =$
{∅, {Niko}, {Sam}, {Frank}, {Luka},
{Niko, Sam}, {Niko, Frank}, {Niko, Luka},
{Sam, Frank}, {Sam, Luka}, {Frank,Luka},
{Niko, Sam, Frank}, {Niko, Sam, Luka},
{Niko, Frank, Luka}, {Sam, Frank, Luka},
{Niko, Frank, Sam, Luka}}.

We can see by counting the number of elements in our power set that there are 16 possible ways that our friends could chat (or not). □

Is there a way we could have answered the question in Example 1.28 without listing out all the elements in the power set? Notice in Example 1.26 the cardinality of the original set is 2 and the cardinality of its power set is 4; for Example 1.27 the cardinality of the original set is 4 as is that of Example 1.28 and the cardinality of each of their respective power sets is 16. This might lead you to suspect that in general, the cardinality of an arbitrary set A is given by the formula

$$|\mathscr{P}(A)| = 2^{|A|}.$$

This formula is actually correct and will work for any set (feel free to show that it works if A has 3 elements). Thus, if a set has 10 elements, then its power set would have $2^{10} = 1{,}024$ elements. Chapter 6 on counting will clarify why this formula works in general.

Remark. In case you were wondering why we call $\mathscr{P}(A)$ the "power set" of A: Some people use the notation 2^A for the power set of A. (This is an instance of a more general notation, which we won't get into here.) Using this alternative notation, we have the following handy formula

$$|2^A| = 2^{|A|},$$

which is pretty easy to remember: the cardinality of the power set is 2 raised to the *power* of the base set. □

1.5 Principle of Inclusion/Exclusion

The principle of inclusion/exclusion is used to calculate the cardinality (i.e., the number of elements) of the union of two or more sets. At first blush, you might guess that the cardinality of the union of two sets is given by the sum of the cardinalities of the individual sets. To see that this is not the case, consider the following scenarios:

- *Scenario 1*: Let

  $$A = \{a, e, i, o, u\} \qquad \text{and} \qquad B = \{b, c, d\}.$$

  Then

  $$A \cup B = \{a, b, c, d, e, i, o, u\},$$

  and so

  $$|A \cup B| = 8 = |A| + |B|.$$

- *Scenario 2*: Now let

  $$C = \{a, e, i, o, u\} \qquad \text{and} \qquad D = \{a, e, i\}.$$

Then

$$C \cup D = \{a, e, i, o, u\},$$

and so

$$|C \cup D| = 5 \neq |C| + |D|.$$

- *Scenario 3*: Finally, let[7]

$$L = \{f, a, c, e\} \qquad \text{and} \qquad S = \{e, g, b, d, f\}.$$

Then

$$L \cup S = \{a, b, c, d, e, f, g\},$$

and so (once again)

$$|L \cup S| = 7 \neq |L| + |S|.$$

The difference between the first scenario (on the one hand) and the second and third scenarios (on the other hand) may be traced to the fact that a set cannot contain repeated elements. Therefore, when we compute the union of two or more sets that have elements in common, the duplicate elements must be removed so as to not double-count them. Given two sets A and B the *principle of inclusion/exclusion* states that

$$|A \cup B| = |A| + |B| - |A \cap B|. \tag{1.2}$$

As the name implies, this formula tells us that some elements are included and some elements are excluded. Were we to include all the elements in sets A and B, the elements that are common to both would be included twice. Hence to get an accurate count, we must exclude one copy of each common element; that is, we must exclude the set $A \cap B$. Once again, this formula tells us that if we want to calculate the cardinality of $A \cup B$, we include all the elements in set A and include all the elements in set B, but exclude one set of all the elements common to sets A and B to avoid counting them twice.

This can be more clearly shown using a Venn diagram that represents the union of two sets A and B. Figure 1.6 clearly shows that $A \cup B$ is represented by the entire shaded area of the interconnected circles. However, it is important to notice that the elements that sets A and B have in common (i.e. $A \cap B$) are represented by the darker area in the middle. To understand

[7] If you're a musician, you'll understand why we chose L and S as the names of these two sets.

this principle it is important to recognize that for any two sets A and B, the three sets $A - B$, $B - A$ and $A \cap B$ are mutually disjoint. For example, if $x \in A - B$ then $x \notin B - A$. Similarly, if $x \in B - A$, then $x \notin A \cap B$. (We'll let you figure out the other four implications.)

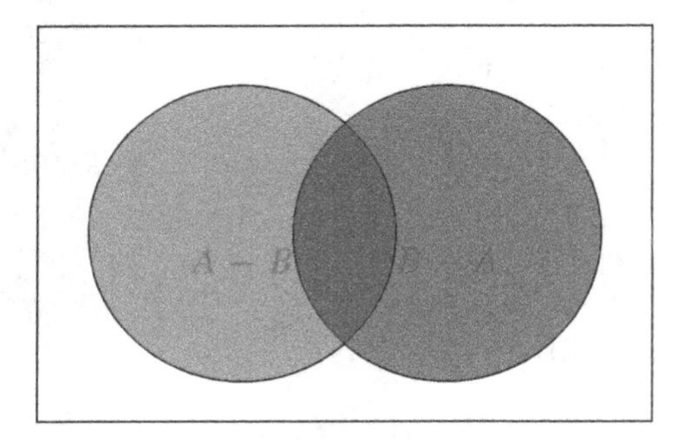

Figure 1.6: $A \cup B$ is represented by the entire shaded area

Example 1.29: A teacher wants to know whether more students have attended Fenway Park or Yankee Stadium (or vice versa). Of the 28 students in the class, 23 students have visited one or both of the stadiums. Of those 23 students, 12 have visited Fenway Park and 9 have visited Fenway Park and Yankee Stadium. Use the inclusion/exclusion principle to answer the following questions:

(a) How many students have visited Yankee Stadium?

(b) How many students have visited only Fenway Park?

Solution: Let

- S denote the set of all students surveyed,
- F denote the set of all students who had visited Fenway Park, and
- Y denote the set of all students who had visited Yankee Stadium,

so that

$$|F \cup Y| = 23$$
$$|F| = 12$$
$$|F \cap Y| = 9$$

(a) Applying the formula

$$|F \cup Y| = |F| + |Y| - |F \cap Y|,$$

we see that

$$23 = 12 + |Y| - 9$$
$$23 = 3 + |Y|$$
$$20 = |Y|$$

So 20 students have visited Yankee Stadium.

(b) The set $F - Y$ consists of all the students who have visited only Fenway Park, and not Yankee Stadium. Now note that $(F \cap Y) \cup (F - Y) = F$. In other words, the set of students who have visited Fenway Park is comprised of the set of students who have visited both Fenway Park and Yankee Stadium (e.g. $F \cap Y$) and the set of students who have only visited Fenway Park (e.g. $F - Y$.) Moreover, the sets, $F \cap Y$ and $F - Y$ are disjoint (i.e. a student who has visited both parks cannot have only visited one.) This can easily be seen by drawing the Venn diagram. We can, therefore, use the principle of inclusion/exclusion on the two sets $F \cap Y$ and $F - Y$ to see that

$$|F| = |F \cap Y| + |F - Y| - |(F \cap Y) \cap (F - Y)|,$$

which we may rewrite as

$$12 = 9 + |F - Y| - 0.$$

So solving for $|F - Y|$ we find that 3 students have visited only Fenway Park. □

Example 1.30: Not surprisingly, all 18 students in a kindergarten class like indoor or outdoor recess. Of those 18 students, 15 like outdoor recess and 11 like indoor recess. Use the inclusion/exclusion principle to determine how many students like both.

Solution: Let I be the set of students liking indoor recess and O be the set of students liking outdoor recess. Then

$$|I \cup O| = 18$$
$$|I| = 11$$
$$|O| = 15$$

Applying the formula $|I \cup O| = |I| + |O| - |I \cap O|$, we see that

$$\begin{aligned} 18 &= 11 + 15 - |I \cap O| \\ 18 &= 26 - |I \cap O| \\ -8 &= -|I \cap O| \\ |I \cap O| &= 8 \end{aligned}$$

We can therefore conclude that 8 students like both indoor and outdoor recess. □

The principle of inclusion/exclusion can be expanded to determine the number of elements resulting in the union of three sets, giving

$$|A \cup B \cup C| = \\ |A| + |B| + |C| - |A \cap B| - |A \cap C| - |B \cap C| + |A \cap B \cap C|.$$

Notice that the general principle remains the same: some elements are included and some elements are excluded. However, it is slightly more complicated because with three sets, there are four possible intersections, namely, $|A \cap B|$, $|A \cap C|$, $|B \cap C|$ and $|A \cap B \cap C|$. We can see this more clearly using a Venn diagram shown in Figure 1.7 that represents the union of three sets A, B, and C.

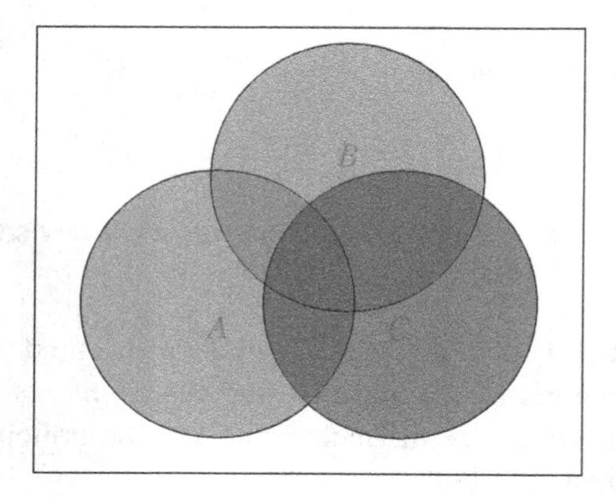

Figure 1.7: $A \cup B \cup C$ is represented by the entire shaded area

Similar to the union of two sets, Figure 1.7 shows that $A \cup B \cup C$ is represented by the entire shaded area. As in the prior examples, when calculating the cardinality of these three sets, you cannot double count the elements in the darker shaded areas. Notice however, when the elements from intersecting areas $A \cap B$, $A \cap C$, and $B \cap C$ are removed, none of the elements common to all three sets remain. This is why you must add back the number of elements in the intersection $A \cap B \cap C$, which represents the elements that are common to all three sets.

Example 1.31: A search firm launched an online survey to determine which methods of communication people were using, the choices being telephone, blogs or email. Everyone who chose to answer the survey had to pick at least one method. Of the 223 respondents, 98 stated they used the telephone, 168 stated they used email, and 69 said they blogged, 48 stated they used the telephone and email, 59 stated they blogged and used email and 11 stated they used all three methods. Use the inclusion/exclusion principle to determine how many people used the telephone and blogged.

Solution: Let T be the set of people that used the telephone, E be the set of people that used email and, B represent the set of people that blogged. Then

$$\begin{aligned}
|T \cup E \cup B| &= 223 \\
|T| &= 98 \\
|E| &= 168 \\
|B| &= 69 \\
|T \cap E| &= 48 \\
|E \cap B| &= 59 \\
|T \cap B| &= ? \\
|T \cap E \cap B| &= 11
\end{aligned}$$

Applying the formula

$$\begin{aligned}
&|T \cup E \cup B| = \\
&\quad |T| + |E| + |B| - |T \cap E| - |E \cap B| - |T \cap B| + |T \cap E \cap B|,
\end{aligned}$$

we have

$$\begin{aligned}
223 &= 98 + 168 + 69 - 48 - 59 - |T \cap B| + 11 \\
223 &= 335 - 107 - |T \cap B| + 11 \\
223 &= 239 - |T \cap B| \\
-16 &= -|T \cap B| \\
|T \cap B| &= 16
\end{aligned}$$

We can therefore conclude that 16 people used the telephone and also blogged. □

1.6 Exercises

1.6.1 Let $T = \{3, 5, 7, 9, 11, 22, 44\}$. Which of the following statements are true?

(a) $7 \in T$ true

(b) $22 \in T$ true

(c) $T \in T$ false

(d) $11 \in T$ true

(e) $\{3, 5, 44\} \in T$ false

(f) $\{3, 5, 44\} \subseteq T$ true

(g) $\{3, 5, 7\} \in T$ false

(h) $\{3, 5, 7\} \subset T$ true

(i) $T \subset T$ false

1.6.2 Give an explicit listing for each of the following sets as defined by the corresponding set builder notation:

(a) $\{x : x \in \mathbb{N} \text{ and } x^2 < 64\}$ $\{1,2,3,4,5,6,7\}$

(b) $\{x \in \mathbb{Z} : x^2 < 64\}$ $\{-7,-6,-5,-4,-3,-2,-1,0,1,2,3,4,5,6,7\}$

(c) $\{3x : x \in \mathbb{Z} \text{ and } x \leq 5\}$ $\{0,3,6,9,12,15,-3,-6,-9,-12,-15,\ldots\}$

(d) $\{3x : x \in \mathbb{N} \text{ and } x \leq 5\}$ $\{0,3,6,9,12,15\}$

(e) $\{x : x = 2y + 3 \text{ and y } \in \{1, 2, 3, 4, 5, 6\}\}$ $\{5,7,9,11,13,15\}$

1.6.3 State the set builder notation that defines the set $A = \{4, 8, 16, 32, 64 \ldots\}$.
$\{x : x = 2y \text{ and } y \in \{2,4,8,16,32,\ldots\}\}$

1.6.4 Given the following sets:

$$A = \{x \mid x \in \mathbb{Z} \text{ and } -4 < x \leq 7\}$$ $-4,-3,-2,-1,0,1,2,3,4,5,6,7$
$$B = \{y \in \mathbb{N} \mid y < 9\}$$ $1,2,3,4,5,6,7,8$
$$C = \{x \mid x = y + 3 \text{ for some } y \in \mathbb{Z} \text{ such that } -6 \leq y \leq 6\}$$ $-3,-2,-1,0,1,2,3,4,5,6,7,8,9$
$$D = \{x \mid x \in \mathbb{Z} \text{ and } 3 \leq x \leq 9 \text{ and x is odd}\}$$ $3,5,7,9$
$$E = \{x \in \mathbb{Z} : -5 \leq x \leq 5\}$$ $-5,-4,-3,-2,-1,0,1,2,3,4,5$
$$F = \{6, 3, 7, 4, 1, 0\}$$

Determine the following:

(a) $(C \cap D) \cup (F \cap E)$ {3, 5, 7, 9, 4, 1, 0}

(b) $(A - C)$ {-4}

(c) $(C - A) \times (F \cap D)$ {{8,3}, {8,7}, {9,3}, {9,7}}

(d) $\mathscr{P}(B \cap D)$ {{3}, {5}, {7}, {3,5}, {3,7}, {7,5}, {3,5,7}, {∅}}

(e) $|(F - D) \cap E|$ {4, 1, 0}

1.6.5 Given the following sets:

$$A = \{0, 2, 4, 6, 8, 10\}$$
$$B = \{1, 3, 5, 7, 9\}$$
$$C = \mathbb{Z}$$
$$D = \emptyset$$
$$E = \{x \mid x \in C \text{ and } x > 0 \text{ and } x \text{ is even}\}$$ {2, 4, 6, 8, 10, ...}
$$F = \{x \in C \mid x > 0 \text{ and } x \text{ is odd}\}$$ {1, 3, 5, 7, 9, ...}
$$G = \{1, 2\}$$

Determine the following:

(a) $A \cap C$ {0, 2, 4, 6, 8, 10}

(b) $B - F$ {∅}

(c) $B - (E \cap F)$ {1, 3, 5, 7, 9}

(d) $(A \cup B) \cap E$ {0, 2, 4, 6, 8, 10}

(e) $(A \cup B) - C$ {∅}

(f) $(B \times G) \cup (B \times G)$ {{0,1}, {2,1}, {4,1}, {6,1}, {8,1}, {10,1}, {2,0}, {2,2}, {2,4}, {2,6}, {2,8}, {2,10}}

(g) $\mathscr{P}(G)$ {{1,2}, {1}, {2}}

(h) $|\mathscr{P}(A)|$ $- 2^6 = 64$

(i) $\mathbb{Z} - ((E \cup F) \cup C \cap D))$ {0, -1, -2, -3, -4, ...}

{4x | x ∈ N and x ≥ 1}

1.6.6 From a group of people asked at random which foreign languages they spoke fluently,

- 32 answered they spoke Spanish fluently,
- 18 answered they spoke Italian fluently and
- 9 answered they spoke Spanish and Italian fluently.

Hint: Use a Venn diagram.

(a) How many people are fluent in Italian but not Spanish?

(b) How many people are fluent in Spanish but not Italian?

(c) How many people are fluent in Spanish or Italian?

Hint: Use the principle of inclusion/exclusion.

1.6.7 A freshman at XYZ University has selected the courses to take during his or her first semester freshman year. Given the following sets:

T = {English Literature, Calculus, Biology, Philosophy}
I = {Philosophy, Journalism, Calculus}
C = {English Literature, History, Philosophy, Calculus, Psychology, Spanish, Biology, Visual Arts}

where T represents the set of courses that the student is currently taking, I represents the set of courses the student is interested in taking and C represents the set of courses available.

Use set operators to represent each of the following statements. In addition, give an explicit listing of the resulting set:

(a) The set of courses the student is taking and is interested in.

(b) The set of courses the student is taking or is interested in.

(c) The set of courses the student is taking but not interested in.

(d) The set of courses the student is interested in and still needs to take.

(e) The set of possible courses the student can choose from for next semester.

1.6.8 Let

$$X = \{\text{hat}\},$$
$$Y = \{\text{hat}, \{\text{gloves}\}, \{\{\text{hat,gloves}\}\}, \{\text{hat}, \{\text{gloves}\}\}\},$$
$$Z = \{\emptyset, \{\{\text{hat}\}\}, \{\text{hat}, \{\text{gloves}\}\}\}.$$

(a) What is $|X| + |Y| + |Z|$?

(b) Does $|X| + |Y| + |Z| = |X \cup Y \cup Z|$?

(c) Is $X \subseteq Y$?

(d) Is $X \in Y$?

(e) Is hat $\subseteq Y$?

(f) Is $\{\{\text{hat}, \{\text{gloves}\}\}\} \subseteq Y$?

(g) Is $X \subseteq Z$?

1.6.9 Explain why $|\mathscr{P}(X)| \neq |X|$ for any finite set X.

1.6.10 Is there a set X such that $|\mathscr{P}(X)| = 3$? If so, describe the set X; if not, explain why.

1.6.11 If the $|X| = 0$, how many elements would be in its power set?

1.6.12 Given two sets X and Y, which will grow faster as $|X|$ and $|Y|$ increase: $\mathscr{P}(X \times Y)$ or $\mathscr{P}(X) \times \mathscr{P}(Y)$? Can they ever be the same? Explain your answers.

1.6.13 Suppose that

$$A = \{c, o, m, p, u, t, e, r, s, i, n\}$$
$$B = \{m, a, t, h, e, i, c, s\}$$
$$C = \{c, h, e, m, i, s, t, r, y\}$$
$$D = \{e, n, g, l, i, s, h, t, r, a, u\}$$
$$U = \{a, b, c, d, e, f, g, h, i, j, k, l, m, n, o, p, q, r, s, t, u, v, w, x, y, z\}$$

Determine the following:

(a) $A \cup B$

(b) $A \cap B$

(c) $A \cap B \cap C$

(d) $(D \cup A) - B$

(e) $(A - B) \cup (B - A)$

(f) $(A - B) \cap (B - A)$

(g) $(A \cap C) \cup (B \cap D)$

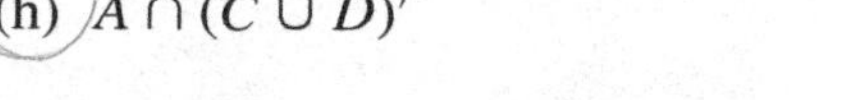

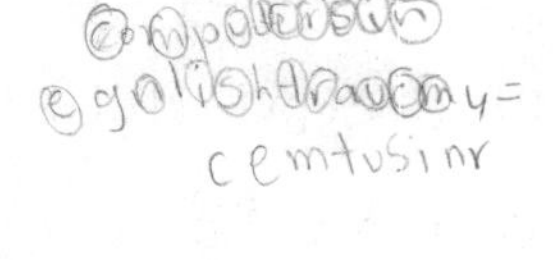

(h) $A \cap (C \cup D)'$

(i) $((A \cup B) \cap (C \cup D))'$

(j) $(A \cup B \cup C \cup D)'$

(k) $|\mathscr{P}(D)|$

(l) $|A \times B|$

1.6.14 Given two sets A and B with $|A| = 9$ and $|B| = 17$. Is it necessarily true that $|A \cup B| = 26$? Explain your answer.

1.6.15 Given two sets X and Y assume that the $|Y| = 20$, $|X \cup Y| = 50$ and $|X \cap Y| = 10$. What is $|X|$?

1.6.16 **[SN]** Suppose that Grace, the most connected user of our social network (on page 317), is creeping around one evening trying to determine who likes what types of music. Of the friends whose profiles she is able to access, she determines that

- Sandra, Alyssa, Lauren and Ellen like classic rock, country, jazz and hip-hop.
- Manny, Chrissy, Niko, and Rania only like jazz.
- Peter likes just classic rock and
- Angela and Luka only like country.

Let

- R be the set of friends who like classic rock,
- C be the set of friends who like country,

- J be the set of friends who like jazz, and
- H be the set of friends who like hip-hop.

Answer the following questions, then answer the last three questions using set operators:

(a) Is $H \subseteq R$?

(b) Is $C \subseteq R$?

(c) How many friends like classic rock and country?

(d) Use set notation to show how many friends like classic rock or jazz?

(e) Is the set notation that represents the how many friends who like classic rock and jazz the same?

(f) Use set notation to represent the set of friends who like jazz, but not classic rock.

1.6.17 Let the universal set U be comprised of the set of all students of a university in New York and let

- S be the set of students receiving academic scholarships,
- L be the set of students receiving student loans, and
- W be the set of students who will be working at least fifteen hours per week during the school year.

Use set operators to represent each of the following. In addition draw a Venn diagram and shade in the appropriate region. Note you can assume that if a student is working, he or she is working at least 15 hours a week.

(a) The set of students who are working and receiving both scholarships and loans.

(b) The set of students who are working, but not receiving any financial assistance.

(c) The set of students who are working or receiving student loans.

(d) The set of students who are working and receiving scholarships, but not loans.

1.6.18 Following the same sets specified in Problem 1.6.17, additionally let

- F be the set of female students of this University.

Use set operators to represent each of the following:

(a) The set of all male students.

(b) The set of all female students who are working at least 15 hours per week.

(c) The set of all male students not receiving any financial assistance.

(d) The set of all female students receiving student loans.

(e) The set of all male students receiving scholarships or loans.

1.6.19 Given some set A and some universal set U, determine the following sets (in terms of A). *Note that the answer does not depend on the specific elements of set A and can be answered in general terms.*

Example: $A \cap A' = \emptyset$

(a) $A - A'$

(b) $(A')'$

(c) $U - A$

(d) $A \cup A'$

(e) $(A \cap A) - (A \cup A)$

1.6.20 Sixteen people responded to an e-mail survey on family pets. Of the 16 who responded, 9 stated they had hamsters, 7 stated they had guinea pigs and 4 stated they had neither hamsters or guinea pigs. Use the principle of inclusion/exclusion to determine how many had both hamsters and guinea pigs.

Hint: Use a Venn diagram.

1.6.21 Let X and Y be sets. Is it possible for $X \subset Y$ and $Y \subset X$? Explain why or why not.

1.6.22 [SN] Using our social network diagram (on page 317) and the set of attributes representing the basic information as listed for each friend in Table A.1 (on page 318), let

$$\begin{aligned} N &= \{\text{attributes for Niko}\}, \\ S &= \{\text{attributes for Sam}\}, \\ F &= \{\text{attributes for Frank}\}, \\ L &= \{\text{attributes for Luka}\}, \\ C &= \{\text{all colleges/universities in the U.S.}\}. \end{aligned}$$

Use set operators to represent each of the following:

(a) All the attributes that these four friends have in common.

(b) How many attributes these four friends have in common.

(c) The set of attributes that Niko and Sam have in common or the set of attributes that Frank and Luka have in common.

(d) The set of attributes that are unique to Niko.

(e) The set of colleges attended by these four friends.

1.6.23 [SN] Using our social network graph as represented (on page 317) and the set of attributes representing the basic information as listed for each friend in Table A.1 (on page 318), let

$$\begin{aligned} X &= \{\text{Larry's friends}\} \\ Y &= \{\text{Peter's friends}\} \\ P &= \{\text{attributes for Peter}\} \\ L &= \{\text{attributes for Larry}\} \\ F &= \{\text{all members of the network}\} \\ C &= \{\text{all cities in the U.S.}\} \end{aligned}$$

Use set operators to represent each of the following:

(a) Set of friends of Larry or Peter.

(b) Set of Larry's friends who are not Peter's friends.

(c) Set of friends in the network that are friends of neither Larry nor Peter. F - (X ∪ Y)

(d) Set of attributes that Peter and Larry have in common. P ∩ L

(e) What must be true if $|L \cup P \cap C| = 1$?

they live in the same city?

Chapter 2

Patterns: Sequences, Summations, Mathematical Induction

> The events in our lives happen in a sequence in time, but in their significance to ourselves they find their own order: the continuous thread of revelation.
>
> Eudora Welty

One of the key characteristics of human intelligence is the ability to identify patterns. Spotting patterns has played a major role in human survival, having allowed our ancestors to determine when to migrate from one area to another, when to go hunting and, more importantly, when to protect themselves in potentially dangerous situations. The ability to spot patterns is no less critical today as it not only guides our interaction with each other, but it also impacts how we make decisions relating to our professional and social environment. Just as choosing the perfect birthday present for your mother improves your standing with her, spotting and then acting upon a social pattern can lead to elevating your own personal or social status. Likewise, spotting a pattern relating to the behavior of consumers in the marketplace can help companies to create and successfully market their products. Should we attribute the phenomenal success of the iPod® to a brilliant idea or to the recognition of a consumer pattern desiring portable music with individualized selection?

The ability to recognize and identify patterns remains an integral component of our society. Today, applications incorporating pattern recognition are far reaching and are used in business, construction, forensics, and cryptography, as well as in the critical study of the human genome. It is therefore important to not only learn how to identify patterns, but also to learn how to represent them. A pattern consists of sets of attributes or components whose order (or sequence) may or may not be relevant to the pattern. In some cases it will suffice to identify the various components that form the pattern. In other cases however, the order of these components is significant in identifying the pattern.

To illustrate, consider how personal information entered into our social network site can be used by businesses to predict certain patterns in consumer behavior. A forty year old professional woman living in the New York City area will likely see more advertisements for anti-aging creams on her social network site than a peer living in Kansas City or a twenty year old woman living in the same New York City area. In this case, it is simply important to recognize the facets that may create the consumer pattern, such as age, professional status and geographic region; however the pattern itself is not bound by a specific order of its components. If we compare this to cracking a secret code, we note that understanding the order of the elements in the code is key to breaking the code. In this case, the components that form the pattern are in a specific sequence and understanding that sequence is critical to identifying the pattern.

The ability to recognize and represent sequential patterns in numerical sequences is the focus of Sections 2.1–2.3 (the summations in Section 2.3 also focus on this). Once we identify the property of a particular sequence we might want to prove that the property exists for all components of that sequence. As we shall see, the use of *mathematical induction*, discussed in Section 2.4, is a powerful proof technique that can be used to rigorously establish the properties of numerical sequences.

2.1 Sequences

A *sequence* is an *ordered* list of objects, referred to as *terms*. Although in this section we focus on numerical sequences, a sequence may be formed from any type of object. A sequence can be infinite or finite. If the sequence is finite, then it will have a *length*, which is simply the number of terms in the sequence. Now suppose that we are interested in some particular sequence. In addition, suppose that we can identify a pattern that determines the sequence. Using this pattern, we can then predict future terms in the sequence. Therefore, our main focus in this section is on identifying and describing these patterns.

As we shall see, it is often easy to identify the pattern of a given sequence, but describing it mathematically may be a bit tricky—at least at first. However, with practice, most students will master this skill.

We begin with an example of a very simple sequence:

$$1, 2, 3, 4, 5, 6, \ldots$$

This sequence is an example of an infinite numerical sequence beginning with the number one. Each term in the numerical sequence is referred to by its its position in the sequence (e.g. *first* term, *second* term, *third* term, etc.). The ellipsis (...) in the sequence indicates that the sequence continues and, in this case, is of infinite length (note that if there was a 10 after the ellipsis, then the sequence would be a finite sequence, with the number 10 as the last term). Most people have no problem identifying the pattern of this sequence and understand that the next two terms in the sequence would be the numbers 7 and 8. This sequence could also be described using words as: "the sequence of integers beginning with the number 1, with each new term being 1 more than the preceding term."

Although describing simple sequences in this fashion may feel comfortable and appear straightforward, this would only be true for simple sequences. Describing complex sequences in this manner would actually get quite cumbersome and complicated. The notation

$$a_1, a_2, \ldots, a_k, \ldots$$

gives us a concise (as well as generic) representation of a sequence. Each element (or term) is represented by a variable that is composed of a lowercase letter followed by a numerical subscript that denotes the position of the element in the sequence. The first term would be read as "a sub 1" and the second as "a sub 2", and so forth. In general, an arbitrary k^{th} term would be read as "a sub k."

Remark. By the way, there is nothing special about the choice of the letter a when denoting a sequence. It is perfectly acceptable to talk about sequences

$$b_1, b_2, \ldots, b_k, \ldots$$

$$c_1, c_2, \ldots, c_k, \ldots$$

$$z_1, z_2, \ldots, z_k, \ldots$$

or even

$$\phi_1, \phi_2, \ldots, \phi_k, \ldots.$$

This allows us to talk about different sequences at the same time. It also allows us to make a mnemonic choice of the letter being used to denote the sequence. □

Example 2.1: Given the sequence $2, 4, 6, 8, 10, \ldots$, what are the values for the terms a_1, a_2, a_3, and a_k when $k = 4$ and $k = 5$? How can we represent the general case of a_k for any position k?

Solution: We have

$$\begin{aligned}
a_1 &= 2, \\
a_2 &= 4, \\
a_3 &= 6, \\
a_k &= 8 \text{ (when } k = 4), \\
a_k &= 10 \text{ (when } k = 5), \text{ and} \\
a_k &= 2k \text{ for the general case (as you will see subsequently).}
\end{aligned}$$

□

Why is representing the general case a_k important? As mentioned earlier, our focus when studying sequences is to identify the pattern in the sequence. A mathematical description of this pattern allows us to succinctly describe said sequence. Moreover, it allows us to determine (or, more importantly, *predict*) any term in the sequence. This ability to predict and extrapolate terms based on existing patterns has very strong real-world applications and is particularly important in fields like cryptography and medical research. In the next section we focus on how to describe sequences by finding and describing the patterns within these sequences.

2.2 Describing Patterns in Sequences

Now that we are familiar with the mathematical notation used to represent sequences, let's focus on how we can we use this notation to describe a sequence based on its pattern. Using the same sequence $2, 4, 6, 8, 10, \ldots$ as in Example 2.1, what do we predict the next term a_6 to be and, more importantly, how can we describe the pattern that led us to our conclusion? Although it may be clear that the next term in this sequence is the number 12, our objective is to describe the general pattern mathematically.

One way to describe this pattern so is to notice that each term in the sequence is 2 more than the prior term. Using this pattern we can rewrite our sequence as follows:

$$\begin{aligned} a_1 &= 2 \\ a_2 &= 4 = a_1 + 2 \\ a_3 &= 6 = a_2 + 2 \\ a_4 &= 8 = a_3 + 2 \\ a_5 &= 10 = a_4 + 2 \end{aligned}$$

Following this pattern we can conclude that the next term a_6 would be 2 more than a_5 which would equal 12. Then $a_7 = 14$, $a_8 = 16$, etc. This is a very natural way to describe this sequence, however, what if we wanted to determine the term a_{10234}? To do so following this methodology would require us to calculate each term preceding it. Clearly this would not be practical for very large sequences. We therefore need the ability to calculate any term in the sequence without having to calculate all the terms prior to it. Since each term in the sequence has a corresponding position, let's see if there is a way to associate the term's position with its value. Using this same sequence of terms, we can see below that each term in the sequence is equivalent to its position multiplied by the number 2.

$$\begin{aligned} a_1 &= 2 = 2 \times 1 \\ a_2 &= 4 = 2 \times 2 \\ a_3 &= 6 = 2 \times 3 \\ a_4 &= 8 = 2 \times 4 \\ a_5 &= 10 = 2 \times 5 \\ a_6 &= 12 = 2 \times 6 \end{aligned}$$

Describing the pattern this way, we can predict that the ninth term a_9 would be 2×9 or 18. Moreover, the term a_{10234} would be 2×10234 or 20468.

We have just described our sequence using two different methods. The first method, called the *recursive method*, described each term in the sequence based on its relationship with the previous term. The second method, called the *closed method*, described each term in the sequence based on its position in the sequence. In other words:

- The *recursive method* relates each term in the sequence to a previous term (or terms) in the sequence.
- The *closed method* directly describes each term in the sequence based on each individual term's position in the sequence. This means that

any given term in the sequence can be determined without referring to any of the other terms.

There are other ways to describe sequences. A sequence can be more simply described without using any formula and just known mathematical terminology. For example, the sequence $1, 2, 3, 4, 5, \ldots$ could be described as "the sequence of integers greater than 0." Also, finite sequences can be described by listing all of their terms, although clearly this is only feasible for short sequences.

Warning: The recursive method tells us how a given term in a sequence depends on the previous term or on certain previous terms. This presents us with a problem namely, how to start the sequence. To be specific, suppose we're dealing with a case in which each term depends only on the previous term (rather than, say, the two previous terms). Once we know the first term, we can determine the second; once we know the second term, we can determine the third, and so forth. But we can't get started until we know the first term. In short, when using the recursive method to specify a sequence, we need two pieces of information: we need to know how each term depends on the previous term, and we also need to know the first term, or more specifically, how the sequence begins. □

Regardless of which method is used to describe the pattern of a sequence, the pattern must first be recognized. In some cases, the pattern may be subtle and difficult to recognize, while in others it may be fairly straightforward. For some sequences, it may sometimes be easier to recognize and describe the pattern using the recursive method while for others, the closed method will be clearer. No doubt there will also be cases when both methods will be equally difficult—it depends on the pattern of the sequence. It may also depend on your individual perspective; for any given sequence, there will be people who have an easier time with the recursive method, and there will be those for whom the closed method is simpler.

We begin with an informal example before providing more details on the recursive and closed methods.

Example 2.2: Given the infinite sequence

$$2, 4, 8, 16, 32, \underline{\quad\quad}, \ldots$$

what is the pattern and what is the next value in the sequence?

Solution: *Recursive Method*
Let's start with the recursive method, which involves determining the value

of a term based on the value of one or more previous terms. For this simple problem we need only consider one previous term. To illustrate, if we look at any term in the above sequence (except the first), we see that it is twice the value of the term directly before it. Therefore, the pattern is that the value of any particular term in the sequence is twice the value of the previous term. That is, $4 = 2 \times 2$, $8 = 2 \times 4$, and so on. Thus the next term in the sequence, the sixth term, would have the value 64 (i.e., $2 \times 32 = 64$).

Let's find a less verbose description of this sequence. First, let's define the pattern between terms a bit more formally. In this case we can do that rather simply: *If we let k denote the position of a given term in the sequence, then its immediate predecessor is in position* $k-1$. Therefore, the value of the kth term is twice the value of the $(k-1)^{\text{st}}$ term. Using the notation introduced earlier, we mathematically express the sequence in our example as $b_k = 2b_{k-1}$. Again, we emphasize that b_k is the value of a given term, which is at position k, and that b_{k-1} is the value of its immediate predecessor, which is at position $k-1$. We are saying that the value of a given term is twice the value of the term directly preceding it.

There is only one thing missing, which is to specify the value of the first term of the sequence. Remember that when describing a sequence using the recursive method, we must always explicitly specify how the sequence begins. In this case, we simply note that the first term in the sequence is 2, and so $b_1 = 2$. Hence, using the recursive method, we have found that our sequence may be described by saying

$$\begin{aligned} b_1 &= 2, \\ b_k &= 2b_{k-1} \qquad \text{for } k \geq 2. \end{aligned}$$ □

Solution: *Closed Method*

In trying to describe this same sequence using the closed method, the pattern is not as obvious. This time we have to directly compute the value for any term in the sequence, based only on its position in the sequence and not on the values of the previous terms. In other words, we need to map the position of the term to the value of the term at that position:

- The term at position 1 has a value of 2, which we may denote by writing $1 \mapsto 2$.
- The term at position 2 has a value of 4 (i.e., $2 \mapsto 4$).
- The term at position 3 has a value of 8 (i.e., $3 \mapsto 8$).
- The term at position 4 has a value of 16 (i.e., $4 \mapsto 16$).

- The term at position 5 has a value of 32 (i.e., $5 \mapsto 32$).

Let us start with the first term and see if we can determine a way to describe that term relative to its position in the sequence, more specifically, how we can use the position number 1 to express the value 2. We know that $2 \times 1 = 2$. This gives us the desired term, so let us see if it works for the second term: $2 \times 2 = 4$. This still works so let us try it for the third term: $2 \times 3 = 6$. Since this is the wrong answer, we'll have to abandon this "doubling pattern."

Let's try another approach. Looking at the first term, we see that $2^1 = 2$. Does this pattern work for subsequent terms? By applying this formula to positions 2, 3, 4, and 5, we find that this pattern *does* work: the sequence $2^1, 2^2, 2^3, 2^4, 2^5, \ldots$ is equivalent to $2, 4, 8, 16, 32, \ldots$.

You may have recognized right away that all of the terms in this sequence are powers of 2. This sequence is referred to as a *binary sequence*. Some of you may be familiar with this pattern, while others may have needed to go through trial and error to discover it. Whatever the case, once we recognize this sequence, we see that the closed formula

$$b_k = 2^k \qquad \text{for } k \geq 1$$

represents this sequence.

By the way, we chose the letter b to represent this sequence for the following reasons:

- We wanted to emphasize that you can use any letter you like to denote a sequence.
- The first sequence used a, and so b is a natural choice for the second sequence.
- The letter b will remind you that this is a binary sequence. □

Remark. Since a recursive description of a sequence tells us how to find a given term of a sequence in terms of one or more previous terms, we have seen that we need one or more "starting values," in addition to the recursive description itself, if we want the recursive description to uniquely identify the sequence in question. On the other hand, the closed method tells us how to directly determine a given term in the sequence, in terms of its position in the sequence. Thus we do not need such starting values when using the direct method. □

Binary Numbers

We normally represent integers using the decimal number system. For example, the decimal number 97625 is really shorthand for the following *decimal* (or *base-ten*) expansion

$$97625_{10} = 9 \times 10^4 + 7 \times 10^3 + 6 \times 10^2 + 2 \times 10^1 + 5 \times 10^0.$$

(the subscript 10 emphasizing the fact that this is a base-ten expansion). If we wanted to directly represent this base-ten number in computer hardware, we would need five circuits (one each for the one's place, ten's place, and so forth); each circuit would quite complicated, since it would need to have the ability to represent ten different values, one for each of the digits 0, 1, 2, ..., 8, 9.

The key to a simpler hardware representation is to note that just as the decimal number system is based on the decimal sequence 10, 100, 1000, 10000, 100000, ..., we can envision a *binary number system*, which is based on the binary sequence 2, 4, 8, 16, 32, ... If we were to use such a numbering system, then we would only need to know how hardware could represent the *binary digits* (or *bits*), namely, 0 and 1. This can be easily done in several ways. For instance, we might let the absence or presence of electrical current respectively denote 0 or 1; alternatively, we could let a light bulb's being "off" or "on" represent the bits 0 or 1.

To convince ourselves that this binary numbering system will work, we need to show that any integer can be written in terms of powers of 2. We can see from the binary sequence shown above how the integers 2, 4, 8, 16, 32, ... are expressed as a power of 2. Moreover, we can also write 1 as a power of 2, since $1 = 2^0$. What about integers that are not powers of two (3, 5, 6, 7, 9, 10, ...)? Just as we have decimal expansions, we also have binary (or base-two) expansions, such as

$$\begin{aligned}
3 &= 1 \times 2^1 + 1 \times 2^0 \\
5 &= 1 \times 2^2 + 0 \times 2^1 + 1 \times 2^0 \\
6 &= 1 \times 2^2 + 1 \times 2^1 + 0 \times 2^0 \\
7 &= 1 \times 2^2 + 1 \times 2^1 + 1 \times 2^0 \\
9 &= 1 \times 2^3 + 0 \times 2^2 + 0 \times 2^1 + 1 \times 2^0 \\
10 &= 1 \times 2^3 + 0 \times 2^2 + 1 \times 2^1 + 0 \times 2^0,
\end{aligned}$$

and so forth. Notice that in some instances we multiply the term by 0

Binary Numbers (cont.)

and in others by 1. Whenever we multiply the term by 0 we are turning *off* that term, and whenever we multiply the term by 1 we are turning *on* that term. Therefore, if we ignore the powers of 2 in this representation and write each term as a sequence of zeros or ones, depending on if we want the term included or not, then we can rewrite

$$3 = 11_2 \qquad 5 = 101_2 \qquad 6 = 110_2$$
$$7 = 111_2 \qquad 9 = 1001_2 \qquad 10 = 1010_2,$$

and so forth. As a further example, the decimal equivalent of the binary number 10110101_2 is

$$10110101_2 = 128 + 32 + 16 + 4 + 1 = 181_{10}.$$

Finally, note that both the decimal and binary numbering systems are *positional numbering system*; the former gives meaning a string of decimal digits, and the latter gives meaning to a string of binary digits.

The example we just completed was fairly straightforward and easily described with both methods. However, note that if you were having trouble coming up with the closed formula, you could use the pattern from the recursive formula as a hint. The recursive formula showed us that the values are generated by doubling the previous value. Repeatedly multiplying by the value 2 yields the sequence

$$2, 2 \times 2 = 2^2, 2 \times 2^2 = 2^3, 2 \times 2^3 = 2^4, \ldots .$$

Thus sometimes if you solve the problem with one method, it can serve as a hint for solving it using the other method.

Example 2.3: Describe the sequence

$$3, 5, 7, 9, 11, \ldots$$

using the recursive method and the closed method.

Solution: *Recursive Method*
We start with the recursive method. What is the relationship between any two successive terms? The relationship should be clear: each pair of terms differs by 2. Thus we can use the recursive method to describe the sequence

(which we denote $c_1, c_2, \ldots$), finding that

$$c_1 = 3$$
$$c_i = c_{i-1} + 2 \qquad \text{for } i \geq 2.$$

Again, it is important to note that this formula would not be complete or correct if we did not specify the value of the first term explicitly. □

Solution: *Closed Method*
The closed formula is actually quite difficult if one has not seen a problem such as this before. The trick is to realize that the recursive method has already told us that we can generate successive terms by adding 2. Thus we should be able to get directly to a value by adding the value 2 multiple times, which is the same as multiplying by 2. If we try the formula $c_j = 2j$, we could get the sequence 2, 4, 6, 8, 10, ..., which is not correct; in fact, $2j$ is the closed form expression for a_j from Example 2.1. But often in cases like these the sequence we get is close to the one we want. A quick look shows us that we are just off by 1 each time. If you like, we can even make a table:

j	1	2	3	4	5
$a_j = 2j$	2	4	6	8	10
c_j	3	5	7	9	11

Thus, we see that

$$c_j = a_j + 1 \qquad \text{for } j \geq 1.$$

Since $a_j = 2j$, this gives

$$c_j = 2j + 1 \qquad \text{for } j \geq 1$$

for the closed formula. □

Remark. Note that we have used different letters (i, j, k) as subscripts in these examples. You can use any letter you want for the subscript, although the letters i, j, k, l, m, n are somewhat traditional. The main thing is to be consistent. If you use a particular letter as your subscript on the left-hand side of a recursive rule or closed formula, you must use the same letter on the right-hand side. Thus in Example 2.3, we could write

$$c_n = 2n + 1 \qquad \text{for } n \geq 1$$

but not (e.g.)

$$c_i = 2n + 1 \qquad \text{for } ? \geq 1$$

(note that we couldn't even decide whether to use i or n in the "for" part in this erroneous formula). □

Example 2.4: Describe the sequence

$$1, 2, 6, 24, 120, \ldots$$

using the recursive and closed methods.

Solution: *Recursive Method*

We begin by looking for a pattern for the recursive method. We need to go from 1 to 2, 2 to 6, 6 to 24 and 24 to 120. How do we do this? The obvious ways to get from 1 to 2 is to either add 1 or multiply by 2. The obvious way to get from 2 to 6 is to add 4 or multiply by 3. The obvious way to get from 6 to 24 is to add 18 or multiply by 4. If we wish to continue generating terms in this sequence, we'll have an easier time of it if we choose the simpler pattern. In this case, multiplication provides a simpler pattern than addition, because the numbers needed for multiplication are easier to predict than those needed for addition. After all, we first multiply by 2, then by 3, and then by 4. To check this pattern, will multiplying by 5 take us from 24 to 120? Yes, it will. Note that to get the second term we multiply by 2, to get the third term we multiply by 3, etc. In other words, the value we are multiplying by to calculate each term is actually the position of the term itself. If we let $d_1, d_2, \ldots$ denote our sequence, we see that our recursive formula takes the form

$$d_1 = 1,$$
$$d_k = k\, d_{k-1} \qquad \text{for } k \geq 2. \qquad \square$$

Solution: *Closed Method*

Now we need to specify the closed formula. This is even harder, but we can use the work from the recursive method as a hint. If we look at how to get the fourth term, we see we had to multiply by 2, then 3, then 4. Thus, we come up with the closed formula

$$d_k = 1 \times 2 \times \ldots k \qquad \text{for } k \geq 1.$$

Let's check this. We get:

$$1, 1 \times 2, 1 \times 2 \times 3, 1 \times 2 \times 3 \times 4, \ldots$$

which equals

$$1, 2, 6, 24, \ldots$$

So, it seems to work; in fact, the formula is correct. Note that each term is actually calculated using a sequence that involves products. As it turns out, there is a name for the product of the first k positive integers, namely,

$$k! = 1 \times 2 \times 3 \times \cdots \times k \qquad \text{for } k \geq 1,$$

read as "k factorial." We will use factorials extensively in Chapters 6 and 7, when studying counting and probability. □

Example 2.5: Describe the sequence

$$2, 3, 7, 25, 121, \ldots$$

using the recursive and closed methods.

Solution: *Recursive Method*

The pattern for this sequence (which we'll denote $e_1, e_2, \ldots$) may not seem obvious at first, but by using the technique shown in Example 2.3, we notice that this sequence is exactly the same as the sequence shown in the prior example, but offset by 1. This can best be shown by comparing the two sequences in the following table:

k	1	2	3	4	5
d_k	1	2	6	24	120
e_k	2	3	7	25	121

So

$$e_k = d_k + 1 \qquad \text{for } k \geq 1. \tag{2.1}$$

We had determined the recursive formula for the sequence in the previous example to be

$$\begin{aligned} d_1 &= 1, \\ d_k &= k\, d_{k-1} \qquad \text{for } k \geq 2. \end{aligned}$$

Using (2.1) and the recursive formula above, we see that for any $k \geq 2$, we have

$$e_k = d_k + 1 = k\, d_{k-1} + 1. \tag{2.2}$$

Now since $k \geq 2$, we see that $k - 1 \geq 1$, and so it makes sense to talk about d_{k-1} and e_{k-1}. Again using (2.1), but now replacing k by $k - 1$, we have

$$e_{k-1} = d_{k-1} + 1 \qquad \text{and so} \qquad d_{k-1} = e_{k-1} - 1.$$

Substituting this result into (2.2), we see that

$$e_k = k(e_{k-1} + 1) + 1 = k\, e_{k-1} - k + 1 = k\, e_{k-1} - (k - 1).$$

Therefore the recursive equation we need to generate our desired sequence is

$$\begin{aligned} e_1 &= 2, \\ e_k &= k\, e_{k-1} - (k - 1) \qquad \text{for } k \geq 1\,. \end{aligned}$$

□

Closed Method

Applying a similar methodology to generate the closed formula for this sequence, we find that in this case

$$e_k = k! + 1 \qquad \text{for } k \geq 1,$$

exactly generates the sequence we are interested in. □

Example 2.6: Given the closed formula $z_k = 3k + 5$, determine the recursive formula that yields the same sequence.

Solution: The simplest way is to first generate the actual sequence. This is done by calculating the value of z_k at different positions until the pattern becomes clear. For example: $z_1 = 3 \times 1 + 5 = 8$, $z_2 = 3 \times 2 + 5 = 11$, etc. Doing this for the first 5 terms yields

$$8, 11, 14, 17, 20, \ldots .$$

The recursive method says that we should look at successive terms (at least to begin with). Clearly we only need to add 3 to each term, so the recursive formula is

$$\begin{aligned} z_1 &= 8 \\ z_k &= z_{k-1} + 3 . \qquad \text{for } k \geq 2 \end{aligned}$$

Note that adding the 3 corresponds to multiplying by 3 in the closed formula. The fact that we added 5 in the closed formula only impacts the value for the initial term in the sequence. This should make sense, since adding a constant just shifts the entire sequence by that constant amount. □

2.3 Summations

A *summation* is the sum of the terms in a finite sequence. Thus, corresponding to the sequence 1, 2, 3, 4, 5, 6 we have the summation

$$1 + 2 + 3 + 4 + 5 + 6$$

Of course, the value of this summation is 21. The summation given above is a particular instance of the more general summation

$$1 + 2 + 3 + \cdots + n = \sum_{i=1}^{n} i, \tag{2.3}$$

with $n = 6$.

The expression found on the right-hand side of (2.3) is an example of *sigma notation* (sometimes called *summation notation*), which uses a large version of the Greek letter Σ (pronounced "sigma")[1] to represent sums. This notation contains three key pieces of information required to compute the summation:

- The $i = 1$ at the bottom of the Σ tells us two things:
  - what we will be using as an *index variable* (sometimes called a *variable of summation*), and
  - the starting value of the index variable.

  In this case, the index variable is i, and its starting value is 1.

- The value at the top of the Σ specifies the final value for the index variable. Sometimes this is a fixed integer. In our case, this final value is another variable, namely, n. Since there are n integers in the set $\{1, 2, \ldots, n\}$, we see that there will be n terms in the summation.

- Just to the right of the Σ, you will find an expression, which the summation is telling you to add. This summation expression will generally involve the index variable. In our case, this expression is simply i itself.

Note that the index variable i is incremented by 1 with each successive term summed and the summation continues until $i = n$. Also note that if the expression being summed is stated in terms of the index variable, then the value of the expression will generally change each time that the variable of the summation is incremented.

In this example, the variable $i = 1$ tells us that we begin with the first term and continue to perform the summation of i until (and including) $i = n$. The summation on the right-hand side of the equal sign in equation (2.3), which is expressed using sigma notation, can be read as "the summation from $i = 1$ to n of i." Thus, if n has the value 5, the summation would evaluate to $1 + 2 + 3 + 4 + 5 = 15$. The following examples will give you practice with summations.

[1]The letter Σ is pronounced as an "s" in Greek. This "s" should remind you of the word "summation."

Example 2.7: What are the values given by the summations in the formulas

$$\sum_{i=2}^{8}(i+3) \qquad \text{and} \qquad \sum_{j=3}^{5}(j^3+2)\,?$$

Solution: The first summation is generated using the sequence based on the expression $i+2$. It starts with $i=2$ and ends with $i=8$. So we get

$$\begin{aligned}\sum_{i=2}^{8}(i+3) = (2+3)+(3+3)+(4+3)+(5+3)+(6+3)+ \\ (7+3)+(8+3) \\ = 5+6+7+8+9+10+11 \\ = 56.\end{aligned}$$

The second summation asks us to add up the expression j^3+2, starting with $j=3$ and ending with $j=5$. So we get

$$\begin{aligned}\sum_{j=3}^{5}(j^3+2) &= (3^3+2)+(4^3+2)+(5^3+2) \\ &= 29+66+127 \\ &= 222.\end{aligned}$$

Note that the use of parentheses for each of these examples is very important. By way of contrast, we find that if we omit the parentheses in the first example, we get

$$\sum_{i=2}^{8} i+3 = (2+3+4+5+6+7+8)+3 = 38. \qquad \square$$

Example 2.8: Compute the values of the summations given by the sigma notation,

$$\sum_{i=1}^{4}(3i-2) \qquad \text{and} \qquad \sum_{j=1}^{4} 3j-2.$$

Hint: How does the presence or absence of parenthesis effect the summation?

Solution: In the first summation, the expression is within parenthesis, therefore, the expression $3i - 2$ is summed with each iteration, giving us

$$\begin{aligned}\sum_{i=1}^{4}(3i - 2) &= (3 \times 1 - 2) + (3 \times 2 - 2) + (3 \times 3 - 2) + (3 \times 4 - 2)\\ &= 1 + 4 + 7 + 10\\ &= 22.\end{aligned}$$

In the second summation, there are no parenthesis, therefore, the expression being summed is simply $3j$ and the 2 is subtracted once the summation is complete as,

$$\begin{aligned}\sum_{j=1}^{4} 3j - 2 &= [(3 \times 1) + (3 \times 2) + (3 \times 3) + (3 \times 4)] - 2\\ &= (3 + 6 + 9 + 12) - 2\\ &= 28.\end{aligned}$$

□

Example 2.9: Compute the value of the summation given by the sigma notation,

$$\sum_{i=1}^{6} 9$$

Hint: Notice that the expression being summed is the constant value 9.

Solution: As the expression being summed is not dependent on the index variable i, the expression remains constant through each iteration, giving us

$$\begin{aligned}\sum_{i=1}^{6} 9 &= 9 + 9 + 9 + 9 + 9 + 9\\ &= 54.\end{aligned}$$

□

Example 2.10: Translate the following two sums into sigma notation:

- $5 + 10 + 15 + 20 + 25$
- $1 + 8 + 27 + 64 + 125 + 216$

Hint: The key is to identify the summation expression. This can be done by noticing that these summations represent the summing of a sequence. Therefore, we can use the methods we learned in Section 2.1 to generate the closed formula that describes each sequence. The desired summation expression is then the closed formula expressed in terms of the index variable of the summation.

Solution: For the first summation, 5 is added between each successive term, which tells us that the recursive formula would add 5. This tells us that the closed formula will involve multiplication by 5. In fact, the correct closed formula for the expression being added is $5k$. So we may write the sum as

$$\begin{aligned}\sum_{k=1}^{5} 5k &= 5 \times 1 + 5 \times 2 + 5 \times 3 + 5 \times 4 + 5 \times 5 \\ &= 5 + 10 + 15 + 20 + 25\end{aligned}$$

The second summation is more difficult. There are a few tricks that one could use to analyze the terms in the sum, but they are more complicated than simply thinking about the terms and hoping for some insight. After a while, you will perhaps realize that the values of the terms in this sequence are all perfect cubes: $1 = 1^3, 8 = 2^3, 27 = 3^3, 64 = 4^3, 125 = 5^3$, and $216 = 6^3$. Thus the closed formula is $a_k = k^3$. Once we have this, we easily see that the sigma notation has the form

$$\begin{aligned}\sum_{k=1}^{6} k^3 &= 1^3 + 2^3 + 3^3 + 4^3 + 5^3 + 6^3 \\ &= 1 + 8 + 27 + 64 + 125 + 216\end{aligned}$$

□

2.4 Mathematical Induction

Before the advent of MP3 players and hand-held devices, children played simpler games. One of the most popular was the game of dominoes. In one variation of the game, the objective is to lay out the dominoes in such a way that once you tip over the first piece, it will tip over the piece next to it, which in turn will tip over the piece next to it and so forth, triggering a cascading effect that will ultimately knock down all the domino pieces. Of course, all this depends on our dominoes being set up properly, so that when a given domino falls over, its neighbor to the right will fall over as well. After playing the game enough times we naturally form the conclusion that no matter how many pieces we set up, if they're set up properly, then tipping the first domino will cause all the dominoes to fall down.

Can we *establish* our conclusion by simply setting up multiple games of dominoes, increasing the number of pieces in each game and showing that all the pieces are knocked over? No, and the reason is that we cannot possibly try infinitely many possibilities; the best we can do is form a conclusion based on some finite set of test runs. No matter how many tests we run, there's always the possibility that our desired conclusion will not hold for one of the tests we didn't run. Therefore, though we may have formed a valid conclusion, we have not established or proven it.

Before trying to show that all the dominoes will fall over, let's think about what happens when we play the game with a properly setup set of dominoes.

1. We can knock down the first piece.

2. If any one piece is knocked down, it will knock down the piece directly to its right.

We can now show that all the dominoes will fall over. Here's why:

1. We already know that domino #1 will fall over.

2. We claim that domino #2 will fall over. To see this, note that the condition 2 given above tells us that if domino #1 is knocked down, then domino #2 will also fall over. *But we already know that domino #1 will fall over from the previous step.* So domino #2 will fall over.

3. We claim that domino #3 will fall over. To see this, note that the condition 2 given above tells us that if domino #2 is knocked down, then domino #3 will also fall over. *But we already know that domino #2 will fall over from the previous step.* So domino #3 will fall over.

4. We claim that domino #4 will fall over. To see this, note that the condition 2 given above tells us that if domino #3 is knocked down, then domino #4 will also fall over. *But we already know that domino #1 will fall over from the previous step.* So domino #4 will fall over.

Continuing in this manner, it is evident that all the dominoes will eventually fall over.

The proof technique we just employed is an example of mathematical induction, a technique used to show that a given statement is true for all positive integers. It is important to note that we are using this technique to *prove* or *establish* the statement that is supposed to be true for all positive integers, and not to *identify* the statement. This is because (paradoxically enough) mathematical induction is based on *deductive* reasoning, a technique used to prove a conclusion that has already been formed. Fans of legendary detective Sherlock Holmes know that Mr. Holmes used his keen powers of observation to solve his cases first by forming a conclusion (which he kept to himself), and secondly by identifying those facts that supported and thereby proved his conclusion. In technical terms, Mr. Holmes used *inductive* reasoning to form his conclusion and *deductive* reasoning to prove his conclusion. This distinction is important because inductive reasoning can only help us form our conclusion, while deductive reasoning is the method by which we prove it.

2.4.1 First Principle of Mathematical Induction

Analogous to our domino example, using mathematical induction to prove a stated property on the set of positive integers requires us to do two things:

1. First, show that the given property is true for the base case, In other words, show that it is true about the positive integer 1. This is analogous to saying that we will push over the first domino.

2. Second, show that if the property is true for some arbitrary positive integer k, it will also be true for the next element $k+1$ (also called the *successor* of k). This is analogous to saying that if we push over any given domino, it will knock over the neighboring domino directly to its right.

If we can do this, then the property holds for all positive integers. Here is a more concise definition of what we're talking about:

Fact (First[2] Principle of Mathematical Induction For Positive Integers)**.** *Let $P(n)$ be a statement about the positive integer n. Suppose that we can prove the following:*

- ***Basis step:*** *$P(1)$ is true.*

- ***Induction step:*** *If $P(k)$ is true for some arbitrary positive integer k, then $P(k+1)$ is true.*

Then $P(n)$ is true for any positive integer n. □

To understand how mathematical induction establishes that $P(n)$ is true for any positive integer n, we must first understand how we the induction

[2]The fact that we refer to a *first* principle of mathematical induction may lead you to guess that there's a second principle of mathematical induction. That is correct. However, we do not cover it in this book.

step works. We start by *assuming* that $P(k)$ is true; this assumption is called the *inductive hypothesis.* We then use this assumption to establish the truth of $P(k+1)$.

Once we have done both the basis step and the induction step, we then immediately know that $P(n)$ is true for *any* positive integer n. The explanation for this mirrors the explanation of why all the dominoes will fall down, which we saw on page 61:

1. We know that $P(1)$ is true from the basis step.
2. We claim that $P(2)$ true. To see this, note that the induction step tells us that if $P(1)$ is true, then $P(2)$ is true. *But we already know that* $P(1)$ *is true from the previous step.* So $P(2)$ is true.
3. We claim that $P(3)$ true. To see this, note that the induction step tells us that if $P(2)$ is true, then $P(3)$ is true. *But we already know that* $P(2)$ *is true from the previous step.* So $P(3)$ is true.
4. We claim that $P(4)$ true. To see this, note that the induction step tells us that if $P(3)$ is true, then $P(4)$ is true. *But we already know that* $P(3)$ *is true from the previous step.* So $P(4)$ is true.

And so it goes. Starting from the basis step, the inductive step allows us to reach any level after a finite number of steps. So the statement $P(n)$ is true for any positive integer n.

Although mathematical induction may appear confusing at first, it is actually quite straightforward once you understand the following ingredients of the proof:

- *Statement to prove:* A statement $P(n)$ about an arbitrary positive integer n.
- *Basis step:* This is $P(1)$, i.e., $P(n)$ when $n = 1$. This is a direct proof that $P(n)$ is true when $n = 1$.
- *Inductive step:* We need to show that if $P(k)$ is true for an arbitrary positive integer k, then $P(k+1)$ is true.

Students generally find the inductive step to be the hardest. One way to look at this is to consider $P(n)$ to be a template or a stencil, which is different for each different value of n.

1. Write down $P(k)$, which is simply $P(n)$ with each instance of n replaced by k.

2. Now write down $P(k+1)$, which is simply $P(n)$ with each instance of n replaced by $k+1$.

You can think of these as being done via a word-processor's "search and replace" operation. *This means that these operations are completely mechanical; you get correct statements of $P(k)$ and $P(k+1)$ by simply following the search-and-replace instructions.* Then $P(k)$ from step 1 above is a working hypothesis, called the *inductive hypothesis*, that you'll use in proving the desired conclusion in step 2.

2.4.2 Examples Using Mathematical Induction

As always, a couple of examples will help you to understand how this all works. It's traditional that the first induction problem most students see is that of verifying the sum of the first n positive integers; it's also traditional to relate that when the nineteenth-century mathematician Carl Friedrich Gauss was a schoolchild, he discovered this formula on his own.[3]

Example 2.11: Let's show that

$$\sum_{i=1}^{n} i = \tfrac{1}{2}n(n+1) \qquad \text{for all } n \in \mathbb{Z}^+, \tag{2.4}$$

recalling that

$$\sum_{i=1}^{n} i = 1+2+3+4+\cdots+n.$$

Solution: To set up our proof by mathematical induction, we must show that:

(a) $P(1)$ is true (basis step), and

(b) for any $k \in \mathbb{Z}^+$, the truth of $P(k)$ implies the truth of $P(k+1)$ (induction step).

Here, $P(n)$ is the statement

$$\sum_{i=1}^{n} i = \tfrac{1}{2}n(n+1). \tag{2.5}$$

[3] See `http://www.sigmaxi.org/amscionline/gauss-snippets.html`, which has a collection of 109 different anecdotes that tell this story.

Basis Step: To establish $P(1)$, note that we want to prove that

$$\sum_{i=1}^{1} i = \tfrac{1}{2} \cdot 1 \cdot (1+1). \tag{2.6}$$

Now the left-hand side of (2.6) is 1, since it consists of a single term beginning and ending at 1, which is simply 1, whereas the right hand side evaluates as,

$$\tfrac{1}{2} \cdot 1 \cdot (1+1) = \tfrac{1}{2} \cdot 1 \cdot 2 = 1.$$

So the left-hand and right-hand sides of (2.6) both equal 1. This completes the basis step.

Inductive Step: We need to show that if $P(k)$ is true for some $k \in \mathbb{Z}^+$, then $P(k+1)$ is true.

> **Note:** Before we continue with proving the inductive step, let us review a basic property of summations. Suppose that we wanted to compute $1+2+3+\cdots+10+11$. One way to do this would be to simply add the numbers from 1 to 11. But suppose that we already knew that $1+2+3+\cdots+10 = 55$. We could use this knowledge to give us a quicker way of evaluating the desired sum, by substituting 55 for $1+2+3+\cdots+10$ in the expression $1+2+3+\cdots+10+11$. In more detail, we have
>
> $$\begin{aligned} 1+2+3+\cdots+10+11 &= (1+2+3+\cdots+10)+11 \\ &= 55+11 \\ &= 66. \end{aligned}$$
>
> With this in mind let us return to our solution.

Our inductive hypothesis is that $P(k)$ is true for some $k \in \mathbb{Z}^+$, i.e., that

$$\sum_{i=1}^{k} i = \tfrac{1}{2}k(k+1). \tag{2.7}$$

Using this inductive hypothesis, we want to prove that $P(k+1)$ is true. We find $P(k+1)$ by substituting $k+1$ for n in $P(n)$. So we need to prove that

$$\sum_{i=1}^{k+1} i = \tfrac{1}{2}(k+1)\big((k+1)+1\big), \tag{2.8}$$

which we may rewrite as

$$1 + 2 + \cdots + k + (k + 1) = \tfrac{1}{2}(k + 1)(k + 2). \tag{2.9}$$

Note that we obtained (2.7) by simply replacing all instances of n in (2.4) by k, and we obtained (2.8) by simply replacing all instances of n in (2.4) by $k + 1$. (As we said earlier, this is analogous to a search-and-replace operation that you can do in a word processor.) We also used the definition of what summation notation means.

Referring back to (2.9), we know that if the last term in the sum is $k+1$ then the term directly before it must be k. Using the idea given in the note above, let's rewrite the statement (2.9) we want to prove as

$$(1 + 2 + 3 + 4 + \cdots + k) + (k + 1) = \tfrac{1}{2}(k + 1)(k + 2). \tag{2.10}$$

The inductive hypothesis (2.7) tells us that

$$1 + 2 + 3 + 4 + \cdots + k = \tfrac{1}{2}k(k + 1).$$

This means that we can replace $1 + 2 + 3 + 4 + \cdots + k$ by $\frac{1}{2}k(k + 1)$ in (2.10), which gives us the equality

$$(1 + 2 + 3 + 4 + \cdots + k) + (k + 1) = \tfrac{1}{2}k(k + 1) + (k + 1). \tag{2.11}$$

Looking the equation (2.10) we want to prove and the equation (2.11) that we know to be true, we see that we are done if we can prove that the right-hand sides of these two equations are the same. Using some algebra (the distributive law twice, along with the commutative law), we find that

$$\begin{aligned}\tfrac{1}{2}k(k + 1) + (k + 1) &= \tfrac{1}{2}k \cdot (k + 1) + 1 \cdot (k + 1) = (\tfrac{1}{2}k + 1)(k + 1)\\ &= \tfrac{1}{2}(k + 2)(k + 1) = \tfrac{1}{2}(k + 1)(k + 2).\end{aligned}$$

Thus the right-hand sides of (2.11) and (2.10) are the same, which completes the induction step. Having done both the basis step and the inductive step, the first principle of mathematical induction now tells us that $P(n)$ is true for all $n \in \mathbb{Z}^+$, i.e., that

$$\sum_{i=1}^{n} i = \tfrac{1}{2}n(n + 1). \qquad \square$$

For our next example, let us assume that you find an intriguing picture on the Internet one day and upload it onto your computer. You then decide to email the picture to exactly two friends. They each in turn email the picture to exactly two *new* friends, who in turn email the picture to exactly two *new* friends, who in turn email the picture to two exactly two *new* friends, etc. How many *new* friends do you think will receive the picture after the fifth set of email distributions?

There is a very powerful construct in computer science that can model such problems, called a *tree*. (You will see more applications of trees when you study combinatorics in Chapter 6 and graphs in Chapter 9.) Unlike the trees we encounter in nature (which grow bottom-up, with the root at the bottom and the branches going up), trees in computer science grow top-down, with the root of the tree at the top and the branches of the tree growing down. Figure 2.1 shows a typical representation of a tree as used in computer science.

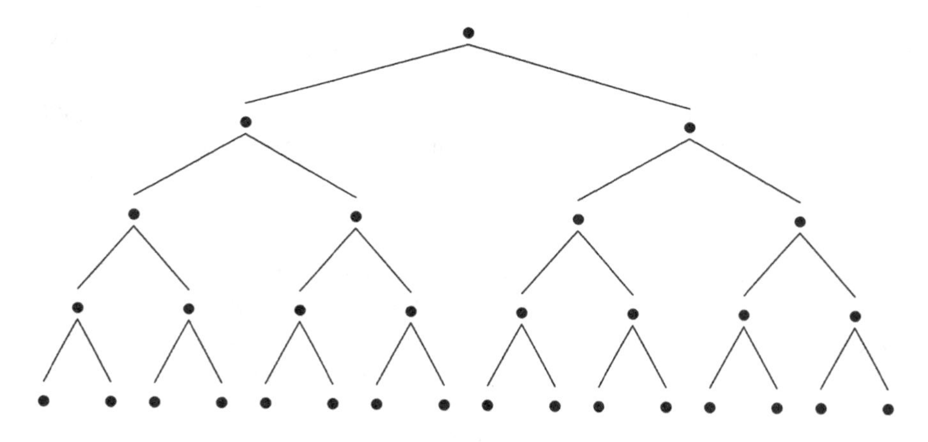

Figure 2.1: Complete Binary Tree

Notice that the tree models our problem in that the tree grows by *doubling* its branches at each *branching point*. This is a representation of a special type of tree known as a *binary tree*;[4] more precisely, this tree is a *complete binary tree*.[5] Each branching point is called a *node*, with the topmost node being called the *root node*. Note that each node in the tree branches to two nodes at the next level, with the exception of the nodes at the bottom row, called *leaves*. Excluding leaves, we can consider each node to be the root of two new complete binary trees, called the *left* and

[4] A tree that has at most two nodes at each branching point.

[5] A binary tree that has *exactly* two nodes at each branching point. Note that some books use different terminology here.

right subtrees rooted at that node.

Remark. You can also see how this kind of structure can arise by imagining this as a variation of the *telephone game*, where one friend tells a secret to two friends and, each of those two friends shares the secret with two new friends who, each in turn share the secret with two new friends, and so forth. If we actually undertook this, one can only imagine what the secret would be after enough levels! If we represented this as a tree structure, the secret will remain precisely preserved, but the tree will grow just as quickly. □

Note that two nodes branch from the first level (i.e., the root) of the tree, four nodes branch from the second level, eight nodes branch from the third level, and sixteen nodes branch from the fourth level. Therefore we answer our original question of "how many new friends will receive the picture after the fifth set of email distributions," by seeing how many nodes will branch from the fifth level.

If we look at the number of nodes at a given level of the complete binary tree in Figure 2.1, you'll see that they form the sequence

$$1, 2, 4, 8, 16.$$

However, we're interested in knowing how many friends receive the picture after a given number of email rounds. After the first round, we have two recipients; after two rounds, we have four; and so forth. So the sequence giving us the number of recipients after one, two, three, or four rounds is

$$2, 4, 8, 16.$$

Some of you may notice that this is part of a numerical sequence that we previously studied in Section 2.1, namely, the binary sequence

$$2, 4, 8, 16, 32, 64, \ldots$$

which grows in powers of 2 as

$$2^1, 2^2, 2^3, 2^4, 2^5, 2^6 \ldots 2^n, \ldots$$

From this we can conclude that at some level n of a binary tree, the number of new nodes that will form will be 2^n. Although we've convinced ourselves of this fact (based on our earlier experience with the binary sequence), this seems to be a good candidate for an induction proof. So, let's give it a try.

Example 2.12: Use mathematical induction to prove that the number of nodes that will branch from the nth level of a complete binary tree is 2^n.

Solution: Following the recipe we've developed, we should let $P(n)$ be the statement about the positive integer n that we want to prove. This means that we should let $P(n)$ be the statement that the number of nodes in the nth level of a complete binary tree is 2^n. Recalling how mathematical induction works, suppose that we can do the following:

(a) Show that $P(1)$ is true (basis step).

(b) Show that for any $k \in \mathbb{Z}^+$, the truth of $P(k)$ implies the truth of $P(k+1)$ (induction step).

Then we will know $P(n)$ is true for all positive integers n.

Basis Step: To establish $P(1)$, we want to show that the number of nodes that form from the 1st level of a complete binary tree is 2^1. We can verify this by seeing that two nodes form from the first level of our binary tree as shown in Figure 2.1.

Inductive Step: Establish that if $P(k)$ is true for some $k \in \mathbb{Z}^+$, then $P(k+1)$ is true. Our *inductive hypothesis* is that $P(k)$ is true for some $k \in \mathbb{Z}^+$, i.e., that the number of nodes that form from the kth level of a complete binary tree is 2^k. Using this inductive hypothesis, we want to prove that $P(k+1)$ is true. Specifically, for our *inductive step* we want to show that the number of nodes that form the $(k+1)^{\text{st}}$ level of a complete binary tree is 2^{k+1}.

Suppose that we have a complete binary tree, and we want to count the number of nodes at its $(k+1)^{\text{st}}$ level. The key observation is that such a tree has two subtrees below its root node, a left subtree and a right subtree. Moreover, the nodes at the $(k+1)^{\text{st}}$ level of the overall tree arise either as k^{th}-level nodes of the left subtree, or as k^{th}-level nodes of the right subtree. By the inductive hypothesis, the kth-level of the left subtree has 2^k nodes, and the same is true of the right subtree. This means that the overall tree will have $2^k + 2^k$ nodes at its $(k+1)^{\text{st}}$ level; the first 2^k nodes arise from the left subtree, and the second 2^k nodes arise from the right subtree. Now

$$2^k + 2^k = 2 \times 2^k = 2^1 \times 2^k = 2^{1+k} = 2^{k+1}. \tag{2.12}$$

So the overall tree has 2^{k+1} nodes at its $(k+1)^{\text{st}}$ level, which is precisely the statement $P(k+1)$ that we needed to prove. This completes the proof of the induction step.

Since we successfully proved both the basis step and the induction step, it follows that $P(n)$ is true for all positive integers n, as was required. □

From this you can see that after the tenth email distribution takes place, $2^{10} = 1{,}024$ new friends will receive your picture. After the twentieth distribution takes place, $2^{20} = 1{,}048{,}576$ friends will receive your picture.

Remark. Of course, what applies to intriguing pictures[6] also applies to jokes, chain letters, and the like. In particular, it also applies to unfounded rumors that are forwarded in chain mails, as such rumors go viral. Such is the power of geometric (or, more commonly *exponential*) growth. We'll discuss this further in Sections 5.7.2 and 6.2.2. □

Example 2.12 showed us how we could determine the number of nodes that formed from any level of a complete binary tree. However, what if we wanted to calculate the total number of nodes of some complete binary tree, including all the levels of that tree? In reference to our original example, we would like to know the total number of people, including yourself, having your picture at any email distribution level.

Figure 2.1 shows that there is 1 node at the root, 2 nodes at the next level, then 4, then 8, and finally 16 nodes at the bottom of the tree. We can therefore calculate the total number of nodes in our tree as

$$1 + 2 + 4 + 8 + 16 = 31.$$

This series can also be expressed as a summation series of the power of 2, as

$$2^0 + 2^1 + 2^2 + 2^3 + 2^4 = 2^5 - 1. \tag{2.13}$$

Now let's think of the general case, in which the complete binary tree has n levels. Following the reasoning given above, we know that the total number of nodes must be $\sum_{i=1}^{n} 2^{i-1}$. Looking at (2.13), we would probably guess that

$$\sum_{i=1}^{n} 2^{i-1} = 2^n - 1. \tag{2.14}$$

This would imply that the total number of nodes of a complete binary tree of some level n is $2^n - 1$. However, we haven't proved this statement. The weak point in our chain of reasoning is that we have only conjectured that (2.14) holds; we haven't proved it. This leads us to our next task:

Example 2.13: Use mathematical induction to prove that (2.14) holds, i.e., that

$$1 + 2 + 4 + \cdots + 2^{n-1} = 2^n - 1$$

[6]Which, believe it or not, some of your friends (or friends of friends, or friends of friends of friends, etc.) may not think are so intriguing.

is true for any $n \in \mathbb{Z}^+$.

Solution: Once again, we want to prove that some statement $P(n)$ is true for all positive integers n. For this example, $P(n)$ is the statement

$$\sum_{i=1}^{n} 2^{i-1} = 2^n - 1$$

or (if you prefer)

$$1 + 2 + 4 + \cdots + 2^{n-1} = 2^n - 1.$$

To give an inductive proof that $P(n)$ holds for all $n \in \mathbb{Z}^+$, we will need to do the following:

(a) Show that $P(1)$ is true (basis step).

(b) Show that for any $k \in \mathbb{Z}^+$, the truth of $P(k)$ implies the truth of $P(k+1)$ (induction step).

Basis Step: To establish $P(1)$ we need to show that

$$\sum_{i=1}^{1} 2^{i-1} = 2^1 - 1. \tag{2.15}$$

Now our summation series only has one term, with $i = 1$. So the left-hand side of (2.15) is $2^{1-1} = 2^0 = 1$, while the right-hand side is $2^1 - 1 = 2 - 1 = 1$. Since $1 = 1$, the left-hand and right-hand sides of (2.15) agree with each other, establishing the basis step. We can further verify this by noting that our tree in Figure 2.1, has one node at the top level of the tree, the root node.

Inductive Step: Establish that if $P(k)$ is true for some $k \in \mathbb{Z}^+$, then $P(k+1)$ is true. Our inductive hypothesis is that $P(k)$ is true for some $k \in \mathbb{Z}^+$, namely, that

$$1 + 2 + 4 + \cdots + 2^{k-1} = 2^k - 1. \tag{2.16}$$

Using our inductive hypothesis, we want to prove that $P(k+1)$ is true, i.e, that

$$1 + 2 + 4 + \cdots + 2^{(k+1)-1} = 2^{k+1} - 1. \tag{2.17}$$

Simplifying the last term in the summation series, we see that our goal is to show that

$$1 + 2 + 4 + \cdots + 2^k = 2^{k+1} - 1. \tag{2.18}$$

Since the last term in the series is 2^k, then the term directly before it is 2^{k-1}, so we can re-write the equation (2.18) we want to prove as

$$1 + 2 + 4 + \cdots + 2^{k-1} + 2^k = 2^{k+1} - 1. \tag{2.19}$$

Our inductive hypothesis tells us that (2.16) is true. So we can replace the sum $1 + 2 + 4 + \cdots + 2^{k-1}$ in (2.19) with $2^k - 1$. Hence we have

$$\begin{aligned} 1 + 2 + 4 + \cdots + 2^{k-1} + 2^k &= 2^k - 1 + 2^k \\ &= 2^k + 2^k - 1 \\ &= 2^{k+1} - 1. \end{aligned}$$

(Note that we used (2.12) to show that $2^k + 2^k = 2^{k+1}$.) Thus $P(k+1)$ is true. We have now completed the induction step.

Having done both the basis step and the inductive step, the first principle of mathematical induction now tells us that $P(n)$ is true for all $n \in \mathbb{Z}^+$, i.e., that

$$\sum_{i=1}^{n} 2^{i-1} = 2^n - 1. \qquad \square$$

So let's count the total number of people who have your picture, including yourself, after a given number of email rounds, assuming that each person sends the picture to two friends. Suppose that there are n rounds of email. Be careful here! Since we're not counting any emails sent out by recipients in the final round of email (i.e., at the last level of the tree), the number of levels of recipients will exceed the number of email rounds by one. In other words, there will be $n + 1$ levels in a tree corresponding to n rounds of email. This means that we will be able to use the result in Example 2.13, but with n replaced by $n + 1$. Thus the total number email recipients will be $2^{n+1} - 1$. So after ten rounds of email, the number of recipients will be $2^{11} - 1 = 2{,}047$; after twenty rounds, the number of recipients will be $2^{21} - 1 = 2{,}097{,}151$.

2.5 Exercises

2.5.1 For each of the following sequences, first determine the next value in the sequence and then determine the recursive and closed formula that generates the sequence.

(a) 2, 4, 6, 8, 10, 12, 14

(b) 16, 64, 256, 1024, ___

(c) 1, 2, 4, 8, 16, 32, ___

(d) 2, 4, 8, 16, 32, 64, ___

(e) 1, 3, 7, 15, 31, 63 ___

Hint: Compare this sequence to the previous sequence.

(f) 2, 5, 8, 11, 14, 17, ___

(g) 10, 20, 30, 40, 50 ___

(h) 7, 7, 7, 7, 7, 7, ___

(i) 10, 100, 1000, 10000, ___

(j) −40, −30, −20, −10, 0, 10, ___

2.5.2 Given the following sequence: 1, 2, 3, 5, 7, 11, 15, 23, 31 ___
Determine the next number in the sequence and state the recursive formula that generates it.

2.5.3 Express each of the following sums using sigma notation:

(a) $5+10+15+20+25+30$

(b) $6+11+16+21+26+31+36$

(c) $1+2+6+24+120+720$

(d) $0+1+8+27+64+125$

(e) $7+7+7+7$

(f) $\frac{2}{3}+\frac{4}{5}+\frac{8}{9}+\frac{16}{17}+\frac{32}{33}$

2.5.4 Express the summation $3+4+5+\cdots+19+20+21+22$ in two different ways, using sigma notation.

2.5.5 Calculate the value of each of the following summations:

(a) $\sum_{h=5}^{9} 4h$

(b) $\sum_{i=1}^{5} 5i$

(c) $\sum_{j=1}^{5} 5$

(d) $\sum_{k=2}^{7} 2^k$

(e) $\sum_{m=2}^{9}(9 - m)$

2.5.6 Calculate the value of each of the following summations:

(a) $\sum_{i=1}^{5}(4i + 3)$

(b) $\sum_{i=1}^{5} 4i + 3$

2.5.7 Use mathematical induction to show that

$$\sum_{i=1}^{n}(2i - 1) = 1 + 3 + 5 + \ldots + (2n - 1) = n^2,$$

i.e., the sum of the first n odd integers is n^2.

2.5.8 Use mathematical induction to show that

$$\sum_{i=1}^{n}(4i - 3) = 1 + 5 + 9 + \ldots + (4n - 3) = n(2n - 1).$$

Chapter 3

Logic

> Contrariwise, if it was so, it might be; and if it were so, it would be; but as it isn't, it ain't. That's logic.
>
> Lewis Carroll,
> *Through the Looking Glass*

You are conversing online with your friends. They have recently seen a really great movie. Or so they say—they are trying to convince you to go and see it. You listen to the points they make, look at the blogs and the recommended sites they mention, and you try to understand how all this relates to your interests, your schedule, your finances and so forth. And then you decide: will you go or won't you go? We make decisions all the time about all kinds of things. The process is so keenly tied up with what we consider "thinking" to be that we are almost unaware of it. Certainly it's not intuitively and immediately clear to many of us that the process of making decisions could be formalized on paper, or formalized in an electromechanical incarnation. However, somewhat surprisingly, this *is* the case. And without it, the social networking infrastructure you use to communicate and to research your options would not be possible.

The previous chapter has given us a vocabulary to describe patterns and understand their growth, and to describe collections of symbols (for example the names of people in your family) and how they can interact with one another. Now we want to understand how, given this vocabulary, we can solve *symbolic reasoning* problems that involve *logic* and *logical operations*. Our social network might include people who like horror films and people who love adventure flicks. However, unless our social network tools understand that both horror films and adventure flicks are kinds of movies, they won't put these collections of people together.

3.1 Propositional Logic

The numerical algebra you learned in high school helped you to solve numerical problems. One part of this problem-solving strategy was to identify important parts of the problem, assigning them to variables, and then using algebraic laws and formulas to solve the problem. Symbolic logic is similar: you identify important parts of a logical argument, assign them to *logical variables*, and then use the laws and formulas of symbolic logic to determine whether or not the logical argument is valid.

So our first step is choosing variables to denote the various basic statements appearing in the problem. Since our end goal is to determine the validity of a logical argument, these basic statements appearing in the argument will be either true or false. Such statements are said to be *propositions*. In other words, a *proposition* is a statement that is either true or false.

Example 3.1: The following statements are propositions:

(a) "A New York City subway fare is $2.25." This is true (at the time this book was written).

(b) "It will rain by the end of the day." This will be either true or false; however, we may not know which until the end of day. Of course if it rains, we'll already know the answer.

(c) "Molly has posted her holiday photos on flickr." This will be either true or false, depending on what Molly did.

(d) "Fred will always accept a Facebook friend request from a friend of a friend." This will be either true or false, depending on Fred's policy in this matter.

(e) "All even numbers are integers." This is true.

(f) "All integers are even numbers." This is false.

(g) "Some integers are even numbers." This is true.

(h) "One of the authors of this book is allergic to chocolate." This is true; chocolate and red wine are migraine triggers for one of us.

(i) "All of the authors of this book are allergic to chocolate." This is false.

The following sentences are *not* propositions:

(a) "Are you cold?" This is a question, not a statement. (However, "You are cold." *is* a proposition.)

(b) "Colorless green ideas sleep furiously." This sentence, composed by Noam Chomsky in 1957, is an example of a syntactically correct statement that is semantically meaningless. In other words, the grammar is fine, but the sentence is meaningless.

(c) "The integer x is an even number." Note that x can have *any* integer value. Since the value of x can vary, we call x a *variable*. It is clear that this statement's truth value depends on the value of the variable x. For example, if $x = 1$, then the statement is false, whereas if $x = 2$, then the statement is true. Since the variable x is allowed to have *any* integer value, the statement (as given) doesn't have a well-determined truth value. What we have here may be thought of as being a "proposition depending on a variable"; such critters are called *predicates*, which we study in Section 3.2.

(d) "$x^2 + y^2 = z^2$." This statement's truth value depends on the value of the variables x, y, and z. The remarks we made in the previous case hold here as well. □

Example 3.2: Consider the sentence

$$\text{This statement (3.1) is false.} \tag{3.1}$$

Is (3.1) a proposition? For it to be a proposition, it must be either true or false. Let's consider the possibilities:

- Suppose that (3.1) is true. Then we're saying that "This statement (3.1) is false." Thus we see that (3.1) is false. This is a contradiction, and so (3.1) cannot be true.

- Suppose that (3.1) is false. Then it is false that "This statement (3.1) is false." In other words, the statement (3.1) is true. This is a contradiction, and so (3.1) cannot be false.

Since (3.1) can be neither true nor false, we see that (3.1) is not a proposition. This state of affairs is called the *liar paradox*. □

Once we have determined the propositions in a particular argument, we will generally represent them by *logical variables*. For example, we might let

$$p = \text{"A New York City subway fare is \$2.25."}$$
$$q = \text{"It will rain by the end of the day."}$$
$$r = \text{"All even numbers are integers."}$$

and so forth.

Warning: In algebra, we sometimes aren't all that careful about distinguishing between a variable and its value. The same is true in logical algebra; we'll often talk about "the proposition p," rather than "the proposition represented by the variable p." □

3.1.1 Logical Operations

> The human mind has never invented a labor-saving machine equal to algebra.
>
> (Author unknown)

Think of a lawyer preparing to defend a client. Not only must she assemble the facts of the case, but she must compose an argument that will convince a jury that her client is not guilty. Similarly, it is not enough to simply identify propositions and assign their values to variables (which is analogous to assembling the facts of the case). Just as we know the basic arithmetic operations that can be performed upon numerical quantities and the variables representing these quantities, we need to know what kinds of operations can be done to logical values and variables. Once we know how these operations work, we can use them in assembling a logical argument. These "logical operations" are often called *Boolean operations*, or the operations of *Boolean algebra*, in honor of George Boole, a nineteenth century mathematician and philosopher, who was the first to systematically study such logical operations.

The simplest operations are *unary*, meaning that they only operate upon one item (or *operand*). The most important unary logical operation is *logical negation*, often called the NOT operation. Logical negation turns true into false and false into true. (Think of this as being analogous to negating a numerical quantity, e.g.,turning 2 into -2.) We denote the (logical) negation of a logical variable p by p'. The value of the NOT operation is given by the following *truth table*:

p	p'
T	F
F	T

In a truth table, we use "T" and "F" to respectively denote "true" and "false," which saves a bit of space. Sometimes people let "1" and "0" represent "true" and "false"; we'll follow this convention when we discuss logic circuits in Section 3.1.10.

The first row of the truth table identifies the variable name p and the value p' of the Boolean operation. The remaining rows tell us how to compute logical negation:

- When $p = \text{T}$, we have $p' = \text{F}$, i.e., $\text{T}' = \text{F}$.
- When $p = \text{F}$, we have $p' = \text{T}$, i.e., $\text{F}' = \text{T}$.

As you can see, truth tables are compact ways of defining logical operations, analogous to the addition and multiplication tables you learned in elementary school.

Example 3.3: Let p be the proposition "Fred is on Facebook." What would p' mean? What about $(p')'$?

Solution: The expression p' means "It is not the case that Fred is on Facebook;" it would be more natural to say that "Fred is not on Facebook." The expression $(p')'$ means "It is not the case that Fred is not on Facebook;" most of us would simply say that "Fred is on Facebook." □

We now move on to *binary* logical operations, the word "binary" meaning that these logical operations have two operands.[1] There are two basic binary operations:

- *Logical conjunction*, also called the AND operation, and denoted by the symbol $\wedge$. The conjunction is true if and only if *both* of its operands are true.
- *Logical disjunction*, also called the (inclusive) OR operation, and denoted by the symbol $\vee$. The disjunction is true if and only if *either* of its operands are true.

We can summarize these two operations in the following truth tables:

[1] Binary operations are sometimes called *connectives*, because they *connect* two operands.

p	q	$p \wedge q$
T	T	T
T	F	F
F	T	F
F	F	F

p	q	$p \vee q$
T	T	T
T	F	T
F	T	T
F	F	F

Example 3.4: Let p be the proposition "George is reading email," and let q be the proposition "George needs to get away from the computer and go outside for a change." What do the following mean?

(a) $p \wedge q$.

(b) $p \vee q$.

(c) $(p') \vee q$.

Solution: Here are how these propositions would be written out in English.

(a) George is reading email and George needs to get away from the computer and go outside for a change.

(b) George is reading email or George needs to get away from the computer and go outside for a change.

(c) Either George is not reading email, or George needs to get away from the computer and go outside for a change. □

Warning: We mentioned that $\vee$ is an *inclusive* OR operation. In other words, $p \vee q$ is true *even* when both p and q are true. This isn't the same "or" as we find in conversational usage, which usually has the connotation "either-or." (Think of the sentence, "It will rain today or I will go to the beach." Most people wouldn't go to the beach in rainy weather.) The latter "or" is called the *exclusive or*, often called the XOR operation and denoted by $\oplus$. The truth table of XOR is given by

p	q	$p \oplus q$
T	T	F
T	F	T
F	T	T
F	F	F

Be careful to distinguish OR and XOR! □

We next look at the *conditional* operation, denoted by $\Rightarrow$, and often referred to as *implication* or even *conditional implication*. The formula $p \Rightarrow q$ may be read in several different ways:

- If p, then q.
- p implies q.
- p only if q.
- p is sufficient for q.
- q is necessary for p.

Determining the truth table for $\Rightarrow$ requires a bit of thought. The main idea is that $p \Rightarrow q$ should capture the notion of "logical implication" as found in common usage, and so this notion should be reflected in the truth table for $p \Rightarrow q$. Now the first two rows of the truth table are fairly self-evident:

p	q	$p \Rightarrow q$
T	T	T
T	F	F
F	T	?
F	F	?

(3.2)

In other words, if we think of the proposition represented by p as our *hypothesis* and the proposition represented by q as our *conclusion*, then obtaining a true conclusion from a true hypothesis is valid, but obtaining a false conclusion from a true hypothesis is invalid. The tricky part is filling in the last two rows. What can we say about the validity of an implication in which the hypothesis is false? You may be surprised to know that such an implication is always valid, i.e., the truth table of the conditional is

p	q	$p \Rightarrow q$
T	T	T
T	F	F
F	T	T
F	F	T

(3.3)

There are a couple of ways to understand why this is the case. The simplest is to say that once we admit a false hypothesis into an argument, then we can derive anything. (In other words, be careful about which hypotheses you accept!) For example, Bertrand Russel, a famous philosopher of the 19th and 20th centuries (who happened to be an atheist) once proved the statement: "If $2 + 2 = 5$, then I am the Pope."[2] A more mathematical ex-

[2]His proof? "Suppose that $2 + 2 = 5$ and that I am *not* the Pope. Since I am not the Pope, the Pope and I are two people. Since $2 + 2 = 5$, we have $4 = 5$. Subtracting 3 from each side of this equation, we find that $1 = 2$. Since the Pope and I are two people and $2 = 1$, the Pope and I are the same person, i.e., I am the Pope."

planation is provided (for those who dare) in the "dangerous bend" section that follows.

The last binary operation we'll look at is the *biconditional* operation, denoted by $\iff$, and often referred to as *biconditional implication*, by way of contrast with the conditional implication operation $\Rightarrow$. The formula $p \iff q$ may be read in several different ways:

- p if and only if q.
- p is necessary and sufficient for q.
- p is logically equivalent to q.

The truth table for $\iff$ is as follows:

p	q	$p \iff q$
T	T	T
T	F	F
F	T	F
F	F	T

Warning—only for the truly bold![3] We can now give another explanation for the last two rows in the truth table for the conditional operator $\Rightarrow$. Simply stated: if we choose any other possible definition for $\Rightarrow$, then we wind up with a (simple) binary operation that we've already defined.

There are four choices for how to fill in the two "question-mark" slots of the truth table (3.2) for $\Rightarrow$ with either a T or an F, namely, (T, T), (T, F), (F, T), and (F, F). The definition of $\Rightarrow$ that we gave in (3.3) uses the choice (T, T). Let's see how the other three choices (which we shall denote $\Rightarrow_1$, $\Rightarrow_2$, and $\Rightarrow_3$) fare. First, let's look at their truth tables (which we have combined, thereby saving a bit of space):

p	q	$p \Rightarrow_1 q$	$p \Rightarrow_2 q$	$p \Rightarrow_3 q$
T	T	T	T	T
T	F	F	F	F
F	T	T	F	F
F	F	F	T	F

Note the following:

- The truth table for $p \Rightarrow_1 q$ is the same as that for q itself.

[3] If you're not truly bold, you can skip this section without penalty.

- The truth table for $p \Rightarrow_2 q$ is the same as that for $p \iff q$.
- The truth table for $p \Rightarrow_3 q$ is the same as that for $p \wedge q$.

So unless we want $\Rightarrow$ to simply mimic another binary operation, the only reasonable way we have to fill in the question marks in (3.2) is given by (3.3). □

Remark. In Section 1.3, we said that for any set S, we always have $\emptyset \subseteq S$, see page 11. The explanation given there only works for finite sets. Here's a justification that works in general.

Recall that $A \subseteq B$ means that $x \in A \Rightarrow x \in B$ must be true. Letting $A = \emptyset$ and $B = S$, we see that $\emptyset \subseteq S$ means that $x \in \emptyset \Rightarrow x \in S$ must be true. The latter is an implication with a false hypothesis, since the empty set contains no elements. But an implication with a false hypothesis is always true; if you need to, go back and look at truth table for the conditional operator to double-check this. Since $x \in \emptyset \Rightarrow x \in S$ is true, we have $\emptyset \subseteq S$. (Whew!) □

3.1.2 Propositional Forms

When you learned arithmetic and algebra, you started out with the basic arithmetic operations—addition, subtraction, multiplication, and division. You then learned how to use these operations to form simple arithmetic expressions, such as $1 + 2$ and $3 - 4$. Moreover, you learned that these simple expressions could be further combined to yield more complicated expressions; for example, we can put a multiplication sign between the expressions $1+2$ and $3-4$, getting the expression $(1+2) \times (3-4)$. Proceeding in this manner, one can build up whatever arithmetic expressions we need.

The same general idea is true for propositions. We can build new (complicated) propositions out of old (simpler) ones. For example, starting with the *Boolean variables* p and q, we can get *Boolean expressions* (sometimes called *propositional forms*) such as p', q', $p \wedge q$, $p \vee q$, $p \wedge (q')$, and so forth. We can further combine these to build Boolean expressions such as $(p \wedge q)'$, $(p \vee q)'$, $(p \wedge (q')) \vee (p')$, and so forth. As you can see, the process can continue indefinitely, allowing us to build up Boolean expressions of arbitrary complexity.

Remark. The description in the preceding paragraph wasn't very precise. We gave examples of Boolean expressions, as well as examples of how complex expressions can be built out of simpler expressions, but we didn't actually precisely define "Boolean expression." This is fairly easy to do:

- Any Boolean variable is a Boolean expression.
- If P is a Boolean expression, then P' is also a Boolean expression.
- If P and Q are Boolean expressions, then so are $P \wedge Q$, $P \vee Q$, $P \oplus Q$, $P \Rightarrow Q$, and $P \iff Q$.
- Nothing else is a Boolean expression.

Note that this is a recursive definition, in the spirit of Section 2.2. □

Warning: Of course, Boolean expressions are not the same as Boolean variables; the former can be (much!) more complicated than the latter. Sometimes we will need to refer to generic (rather than specific) Boolean expressions (as in the preceding Remark). So we'll give you a bit of typographic help here. Whenever we think that it's important to make the distinction, we'll use lower case letters (such as p and q) to denote Boolean variables, whereas capital letters (such as P and Q) will denote (generic) Boolean expressions. □

One thing you'll notice is that when we combine Boolean expressions in this manner, we're starting to accumulate parentheses at an alarming rate. We can make things a little easier by using parentheses of varying sizes, writing things such as

$$\Big(\big(p \wedge (q')\big) \vee (p')\Big) \vee \big((p') \vee q\big).$$

We can also use brackets as an alternative to parentheses, getting

$$\big([p \wedge (q')] \vee (p')\big) \vee [(p') \vee q].$$

But this is still pretty hard to decode.

We can adapt an idea that you've seen before when studying arithmetic and algebra. When you see something like $-1 + 3$, you instinctively know that the minus sign applies to the 1, and not to the whole expression $1 + 3$. Similarly, when you see $2 \times 1 + 7$, you know that the 2 and the 1 are multiplied, with the result being added to the 7; you would *not* take the 2 and multiply by the sum $1 + 7$. Moreover, if you wanted to override this ordering of operations, you'd use parentheses, so that you would write $-(1 + 3)$ or $2 \times (1 + 7)$. In short, we have the following set of *precedence rules*:

1. Parenthesized subexpressions come first.
2. Negation (unary minus) comes next.
3. Multiplicative operations($\times$, $\div$) are done before additive operations ($+$, $-$).
4. Use parentheses if you have *any* doubt.
5. Evaluate ties left-to-right.

An analogous set of precedence rules works for Boolean expressions. The following list is not complete, but it is good enough for the vast majority of examples you'll encounter:

1. Parenthesized subexpressions come first.
2. Negation ($'$) comes next.
3. Multiplicative operation($\wedge$) is done before additive operations ($\vee$, $\oplus$).
4. Use parentheses if you have *any* doubt. Always use parentheses if you have multiple conditionals.
5. Evaluate ties left-to-right.

Example 3.5: Prune out the unnecessary parentheses in the following Boolean expressions.

(a) $(p \wedge q) \vee r$.

(b) $(p \vee q) \wedge r$.

(c) $(p') \wedge q$.

(d) $(p') \vee (q \wedge r)$.

(e) $((p \wedge q) \vee (p \wedge (r'))) \wedge (p \vee r)$.

Solution: The simplified expressions are as follows:

(a) $p \wedge q \vee r$.

(b) $(p \vee q) \wedge r$. (This one can't be simplified.)

(c) $p' \wedge q$.

(d) $p' \vee q \wedge r$.

(e) $(p \wedge q \vee p \wedge r') \wedge (p \vee r)$. □

Remark. How many different logic operations do we really need? We initially described three basic operations ($'$, $\wedge$, and $\vee$). We then discussed the additional operations $\oplus$, $\Rightarrow$, and $\Longleftrightarrow$, defining them via truth tables. Strictly speaking, these additional operations aren't really necessary. It is possible to show that any truth table can be reduced to a sequence of NOT, AND, and OR operations. In other words, the set $\{', \wedge, \vee\}$ is *universal.* Of course, since this set has three elements, we see that a universal set of size three exists. In the Exercises, we will see that there is a one-element universal set. You can't get much smaller than that! □

3.1.3 Parse Trees and the Operator Hierarchy (∗)

We have seen that we can build up propositions of bewildering complexity by proceding step-by-step from simpler propositions. For example, we can build the proposition

$$[(p \vee q) \wedge ((p') \vee r)] \Rightarrow [(p \Longleftrightarrow q) \vee (p \vee r)]$$

in the following steps:

1. p, q, and r are all propositions (since they're variables).
2. So, $p \vee q$ is a proposition, since $\vee$ is a binary operator.
3. Moreover, p' is a proposition, since $'$ is a unary operator.
4. So $(p') \vee r$ is a proposition.
5. From steps (2) and (4), $(p \vee q) \wedge ((p') \vee r)$ is a proposition.
6. $p \Longleftrightarrow q$ is a proposition.
7. $p \vee r$ is a proposition.
8. From steps (6) and (7), $(p \Longleftrightarrow q) \vee (p \vee r)$ is a proposition.
9. Finally, from steps (5) and (8), we see that

$$[(p \vee q) \wedge ((p') \vee r)] \Rightarrow [(p \Longleftrightarrow q) \vee (p \vee r)]$$

is a proposition.

This gets to be pretty verbose. Fortunately, there is a more compact way of showing how a complicated original expression can be broken down into simpler subexpressions (or, alternatively, how simpler subexpressions can be combined to form more complicated expressions): a *parse tree* of the expression. For example, the parse tree of the expression

$$[(p \vee q) \wedge ((p') \vee r)] \Rightarrow [(p \iff q) \vee (p \vee r)]$$

is

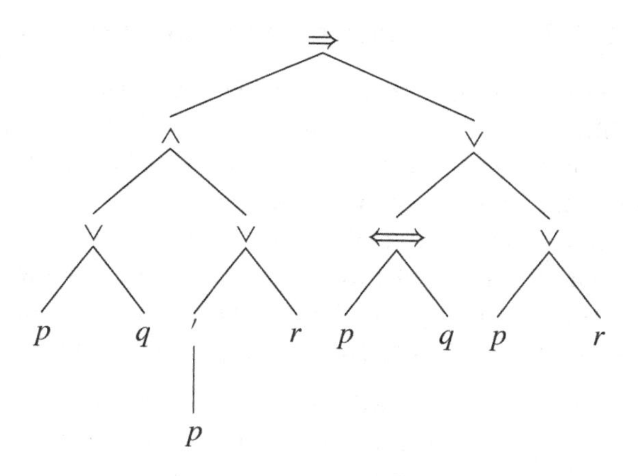

In Section 3.1.2, we showed a simplified set of precedence rules, which allows us to remove extraneous parentheses. However, that set of rules is incomplete. The full set of precedence rules is as follows:

1. Parenthesized subexpressions are evaluated first.

2. Operations have a *precedence hierarchy*:

   (a) Unary negations ($'$) are done first.

   (b) Multiplicative operations ($\wedge$) are done next.

   (c) Additive operations ($\vee$, $\oplus$) are done next.

   (d) The conditional-type operations ($\Rightarrow$ and $\iff$) are done last.

3. In case of a tie (two operations at the same level in the hierarchy), operations are done in a left-to-right order, *except* for the conditional operator $\Rightarrow$, which is done in a right-to-left order. That is, $p \Rightarrow q \Rightarrow r$ is interpreted as $p \Rightarrow (q \Rightarrow r)$.

Following this set of rules, we see that we can write the expression

$$[(p \vee q) \wedge ((p') \vee r)] \Rightarrow [(p \iff q) \vee (p \vee r)]$$

as

$$[(p \vee q) \wedge (p' \vee r)] \Rightarrow [(p \iff q) \vee (p \vee r)]$$

or even

$$(p \vee q) \wedge (p' \vee r) \Rightarrow (p \iff q) \vee p \vee r.$$

3.1.4 Truth Tables, Tautologies, and Contradictions

By now, we've gained some experience constructing Boolean expressions, analogous to the algebraic expressions we learned about in high-school algebra. Let's continue our analogy a bit further. We were often called upon to evaluate algebraic expressions such as x^2, $3x+y$, or a^2+b^2-3abc. That is, we would assign values to the variables occurring in the expression, and then calculate the value of the expression. We are sometimes called upon to do the same thing with Boolean expressions, i.e., determine their values. This is fairly straightforward to do if we use truth tables.

Example 3.6: Find a Boolean expression that would represent the statement "George likes fish, but Freda doesn't like peanut butter," and then construct the truth table for that Boolean expression.

Solution: If we let p denote the statement "George likes fish" and q denote the statement "Freda likes peanut butter," then $p \wedge q'$ means "George likes fish, but Freda doesn't like peanut butter." So let's find the truth table for $p \wedge q'$.

The expression $p \wedge q'$ involves two Boolean variables (p and q). So we'll set up a two-variable truth table for this expression. The rows will begin with all possible truth assignments to the variables p and q, namely (T, T), (T, F), (F, T), and (F, F). We'll start with two columns labeled "p" and "q," under which we'll write these truth assignments. So far, we have

p	q	...?...
T	T	...?...
T	F	...?...
F	T	...?...
F	F	...?...

How should we fill in the remaining columns? We need to calculate all possible values of the Boolean expression $p \wedge q'$, taking into account the truth value of the variables p and q in each case. We could try to do this in our head or on a piece of scrap paper, which is fairly likely to be error-prone. A better idea is to break the expression into its logical pieces, and only calculate the possible values of each piece. So we can evaluate $p \wedge q'$ in two steps:

(a) Calculate q' from the information in the columns to the left of its own column.

(b) Calculate $p \wedge q'$ from the information in the columns to the left of its own column.

The template for our truth table now looks like

p	q	q'	$p \wedge q'$
T	T	?	?
T	F	?	?
F	T	?	?
F	F	?	?

Now we tackle the columns of the truth table, one by one, getting

p	q	q'	$p \wedge q'$
T	T	F	?
T	F	T	?
F	T	F	?
F	F	T	?

and then

p	q	q'	$p \wedge q'$
T	T	F	F
T	F	T	T
F	T	F	F
F	F	T	F

This completes the truth table for $p \wedge q'$. □

Remark. Note that we use double vertical lines to separate the columns that give the values of the variables in questions from the columns that are used to build up the expression. This helps us to keep the truth table well-organized. □

Having whet our blades on a two-variable example, let's try something a bit more challenging: a three-variable example.

Example 3.7: Find a Boolean expression that would represent the statement "Cats are friendly or the telephone is orange; furthermore, the telephone is orange or my sandwich is not moldy." Then construct a truth table for this Boolean expression.

Solution: Let p, q, and r respectively stand for the statements "cats are friendly," "the telephone is orange," and "my sandwich is moldy." Then the Boolean expression $(p \vee q) \wedge (q \vee r')$ means "Cats are friendly or the telephone is orange; furthermore, the telephone is orange or my sandwich is not moldy."

So, we need to find the truth table for $(p \vee q) \wedge (q \vee r')$. Let's follow the ideas in the preceding example. We'll need columns that represent the three variables p, q, and r of the Boolean expression. We'll also need to build up the expression itself in a stepwise manner:

(a) Calculate $p \vee q$ from the information in the columns to the left of its own column.

(b) Calculate r' from the information in the columns to the left of its own column.

(c) Calculate $q \vee r'$ from the information in the columns to the left of its own column.

(d) Calculate $(p \vee q) \wedge (q \vee r')$ from the information in the columns to the left of its own column.

So the template for our truth table will look like

p	q	r	$p \vee q$	r'	$q \vee r'$	$(p \vee q) \wedge (q \vee r')$
T	T	T	?	?	?	?
T	T	F	?	?	?	?
T	F	T	?	?	?	?
T	F	F	?	?	?	?
F	T	T	?	?	?	?
F	T	F	?	?	?	?
F	F	T	?	?	?	?
F	F	F	?	?	?	?

For the time being, don't worry too much about how we managed to figure the truth assignments to the variables p, q, and r, but *do* double-check that all possibilities are covered.[4]

We now fill in the columns, one at a time, as in the previous example. Our final answer will be

[4]The truth values for these three columns don't have to be in the order given; however, unless you establish a methodology for systematically listing the values, you might accidentally omit some of them.

p	q	r	$p \vee q$	r'	$q \vee r'$	$(p \vee q) \wedge (q \vee r')$
T	T	T	T	F	T	T
T	T	F	T	T	T	T
T	F	T	T	F	F	F
T	F	F	T	T	T	T
F	T	T	T	F	T	T
F	T	F	T	T	T	T
F	F	T	F	F	F	F
F	F	F	F	T	T	F

This completes the truth table for $(p \vee q) \wedge (q \vee r')$. □

Remark. In Example 3.6, we saw that a two-variable truth table has four rows. This allowed us to assign all possible values—(T, T), (T, F), (F, T), and (F, F)—to the two variables. Moreover, the values of the two variables were placed at the beginning of each row.

In Example 3.6, we saw that a three-variable truth table has eight rows; once again, each row begins with a truth-value assignment to the three variables in question. If you look closely at the truth table, you'll see that the truth values for the rightmost variable (r) alternate (TFTFTFTF), the truth values for the variable to its left (q) alternate in pairs (TTFFTTFF), and those for the variable to *its* left alternate in blocks of four (TTTTFFFF).

In general, an n-variable truth table has 2^n rows. (This means that nobody would ever want to construct a truth table for a ten-variable proposition, since it would have $2^{10} = 1024$ rows.) The truth values to be assigned to the basic variables follow the same pattern as described above: simple alternation, then alternation of pairs (blocks of size 2), alternation of blocks of size 4, alternation of blocks of size 8, and so forth. □

Note the values of a Boolean expression will generally depend on the values of the variables that the expression uses. This means that generally, a Boolean expression will sometimes be true and sometimes be false. However, some Boolean expressions are always true, whereas some are always false. A Boolean expression that is always true is said to be a *tautology*, whereas a Boolean expression that is always false is said to be a *contradiction*.

Example 3.8: The truth tables of $p \vee p'$ and $p \wedge p'$ are given by

p	p'	$p \vee p'$
T	F	T
F	T	T

and

p	p'	$p \wedge p'$
T	F	F
F	T	F

So $p \vee p'$ is a tautology, and $p \wedge p'$ is a contradiction. This is perfectly sensible. The statement that $p \vee p'$ is a tautology simply means that a variable or its logical complement must always be true; in other words, a variable is either true or false. On the other hand, the statement that $p \wedge p'$ is a contradiction means that it is impossible for a variable and its complement to simultaneously be true; in other words, a variable cannot simultaneously be both true and false. □

3.1.5 Propositional Equivalences

When somebody is trying to convince you that something is true, he may try to use (what he claims to be) a logical argument to make his point. How can we determine the validity of this argument? In the previous section, we saw that statements in a natural language (such as English) can be rewritten as propositions, and that we could use truth tables to determine under which conditions these propositions (and hence the original statements) are true or false. But an argument usually consists of a chain of statements, connected somehow or another. If we want to determine the validity of the argument, we need to know whether all the links in the chain of reasoning are valid. So it is natural to seek general rules that will help us to assess the validity of a logical argument. In this section, we'll discover a number of such rules. Rather than giving examples of how each such rule might work in a particular situation, we'll wait until we've assembled enough rules to do something interesting, see Sections 3.1.8 and 3.1.9

How can we find such rules? Let's do a bit of inductive reasoning, starting with a few of the rules we learned from high-school algebra, such as

$$a + b = b + a,$$
$$a \times (b + c) = a \times b + a \times c,$$
$$-(a + b) = (-a) + (-b),$$

and so forth. You might wonder whether analogous rules hold for propositions. More precisely:

- How would we state such a proposed rule? (Note that we haven't introduced an equal sign for propositions.)
- If such a rule is correct, how would we prove it?
- If it's incorrect, how would we disprove it?

We can easily handle all these issues. Suppose that P and Q are propositional forms. We write $P \equiv Q$ to mean that P and Q are *logically equivalent*, i.e., that P is true if and only if Q is true.

Warning: The statement $P \equiv Q$ is *not* a proposition. It is a statement *about* the propositions P and Q, namely, that the propositions are true (and false) under the same conditions. The operator $\equiv$ is sometimes called a *metasymbol*, and the statement $P \equiv Q$ is a a statement in a *metalanguage* about propositions.[5] The reason for this distinction is that we sometimes need to tell the difference between statements *in* a language (such as the language of propositional forms) and statements *about* that language (e.g., saying that two propositional forms are equivalent). □

Remark. Although $P \equiv Q$ is not a proposition, the statement $P \equiv Q$ holds if and only the proposition $P \iff Q$ is a tautology. □

Looking at the algebraic identities above, we might guess that[6]

$$p \vee q \equiv q \vee p,$$
$$p \wedge (q \vee r) \equiv (p \wedge q) \vee (p \wedge r),$$
$$(p \vee q)' \equiv p' \vee q'.$$

So how would we prove or disprove such conjectured identities?

The answer is simple. We look at the truth table. Suppose we want to prove that $P \equiv Q$ for some (possibly complicated) propositions P and Q. Well, since propositional equivalence means that the two propositions are true (and false) under the same conditions, *the statement $P \equiv Q$ is true if and only if the truth table for P matches the truth table for Q.*

Let's look at the conjectured identities above, and see how they fare.

The first is $p \vee q \equiv q \vee p$. We have

p	q	$p \vee q$
T	T	T
T	F	T
F	T	T
F	F	F

and

p	q	$q \vee p$
T	T	T
T	F	T
F	T	T
F	F	F

[5] By analogy, metahumor consists of jokes about jokes.

[6] Strictly speaking, we don't need parentheses on the right-hand side of the second of these proposed identities. But $p \wedge (q \vee r) \equiv (p \wedge q) \vee (p \wedge r)$ is easier to read than $p \wedge (q \vee r) \equiv p \wedge q \vee p \wedge r$, so we inserted them anyway.

Since the truth values for these two expressions coincide, we see that $p \vee q \equiv q \vee p$, as conjectured. Note that we can combine the two truth tables, which gives us the more concise form

p	q	$p \vee q$	$q \vee p$
T	T	T	T
T	F	T	T
F	T	T	T
F	F	F	F

Since the columns for $p \vee q$ and $q \vee p$ match up (row for row), we see that $p \vee q \equiv q \vee p$, as conjectured.

Remark. Note the use of double vertical lines to separate three parts of the truth table:

- the values of the individual variables,
- the truth table for the left-hand expression in the proposed equivalence,
- the truth table for the right-hand expression in the proposed equivalence.

This is by no means necessary, but it certainly helps keep things well-organized. □

Let's try $p \wedge (q \vee r) \equiv (p \wedge q) \vee (p \wedge r)$. We find

p	q	r	$q \vee r$	$p \wedge (q \vee r)$	$p \wedge q$	$p \wedge r$	$(p \wedge q) \vee (p \wedge r)$
T	T	T	T	T	T	T	T
T	T	F	T	T	T	F	T
T	F	T	T	T	F	T	T
T	F	F	F	F	F	F	F
F	T	T	T	F	F	F	F
F	T	F	F	F	F	F	F
F	F	T	T	F	F	F	F
F	F	F	T	F	F	F	F

Since the columns for $p \wedge (q \vee r)$ $(p \wedge q) \vee (p \wedge r)$ match, we see that $p \wedge (q \vee r) \equiv (p \wedge q) \vee (p \wedge r)$, as conjectured.

Emboldened by our success, let's try the last conjectured identity, namely, $(p \vee q)' \equiv p' \vee q'$. Our truth table

p	q	$p \vee q$	$(p \vee q)'$	p'	q'	$p' \vee q'$
T	T	T	F	F	F	F
T	F	T	**F**	F	T	**T**
F	T	T	**F**	T	F	**T**
F	F	F	T	T	T	T

There's trouble afoot; the columns for $(p \vee q)'$ and $p' \vee q'$ don't match at the second and third lines. This tells us that the two propositional forms are *not* equivalent, which we may write as $(p \vee q)' \not\equiv p' \vee q'$.

It turns out that the correct formula is

$$(p \vee q)' \equiv p' \wedge q'.$$

There's also the companion formula

$$(p \wedge q)' \equiv p' \vee q'.$$

These two formulas are called *De Morgan's laws.* Let's prove the first; the second is similar. We find

p	q	$p \vee q$	$(p \vee q)'$	p'	q'	$p' \wedge q'$
T	T	T	F	F	F	F
T	F	T	F	F	T	F
F	T	T	F	T	F	F
F	F	F	T	T	T	T

Since the columns for $(p \vee q)'$ and $p' \wedge q'$ are identical, we see that

$$(p \vee q)' \equiv p' \wedge q',$$

as conjectured.

3.1.6 Propositional Identities

Some of the identities of propositional logic are useful enough and important enough to have names. A small sampling may be found in Table 3.1.

A few of the identities (e.g., commutative, associative, distributive) in Table 3.1 will come as no big surprise, to the point of appearing somewhat trivial. On the other hand, some of them may not be particularly obvious to you. Regardless of how complicated any of these identities may be, you can prove any of them using truth tables. Alternatively, you can use known identities to prove new identities (just as in high school algebra).

Name	Identity
Identity	$p \wedge T \equiv p$
Identity	$p \vee F \equiv p$
Complement	$p \wedge p' \equiv F$
Complement	$p \vee p' \equiv T$
Double Negation	$(p')' \equiv p$
Idempotent	$p \wedge p \equiv p$
Idempotent	$p \vee p \equiv p$
Commutative	$p \wedge q \equiv q \wedge p$
Commutative	$p \vee q \equiv q \vee p$
Associative	$(p \wedge q) \wedge r \equiv p \wedge (q \wedge r)$
Associative	$(p \vee q) \vee r \equiv p \vee (q \vee r)$
Distributive	$p \wedge (q \vee r) \equiv (p \wedge q) \vee (p \wedge r)$
Distributive	$p \vee (q \wedge r) \equiv (p \vee q) \wedge (p \vee r)$
DeMorgan	$(p \wedge q)' \equiv p' \vee q'$
DeMorgan	$(p \vee q)' \equiv p' \wedge q'$
Modus Ponens	$[(p \Rightarrow q) \wedge p] \Rightarrow q$
Modus Tollens	$[(p \Rightarrow q) \wedge q'] \Rightarrow p'$
Contrapositive	$p \Rightarrow q \equiv q' \Rightarrow p'$
Implication	$p \Rightarrow q \equiv p' \vee q$
Biconditional	$p \iff q \equiv (p' \vee q) \wedge (p \vee q')$
Transitive	$(p \Rightarrow q \wedge q \Rightarrow r) \Rightarrow (p \Rightarrow r)$

Table 3.1: Some handy propositional laws

Example 3.9: Assume that all the identities in Table 3.1 are correct. Use the propositional laws to prove the *exportation identity*

$$[(p \wedge q) \Rightarrow r] \equiv [p \Rightarrow (q \Rightarrow r)].$$

Solution: We have

$$\begin{aligned} (p \wedge q) \Rightarrow r &\equiv (p \wedge q)' \vee r && \text{implication} \\ &\equiv (p' \vee q') \vee r && \text{DeMorgan} \\ &\equiv p' \vee (q' \vee r) && \text{associative} \\ &\equiv p' \vee (q \Rightarrow r) && \text{implication} \\ &\equiv p \Rightarrow (q \Rightarrow r) && \text{implication} \end{aligned}$$

as required. □

Remark. Note the similarity of Tables 1.1 on page 22 and 3.1. Many of the laws have the same (or similar) names; note that such "law pairings" between the tables, the laws are almost typographically the same, with $\cap$, $\cup$, $'$, $\emptyset$, and U being replaced by $\wedge$, $\vee$, $'$, F, and T, respectively. This is no mere coincidence. Had we decided to do logic before we did sets, we could've used the identities in Table 3.1 to establish the analogous identities in Table 1.1.

For example, let's show that the first of DeMorgan's laws in Table 1.1 is true, i.e., we'll show that $(A \cap B)' = A' \cup B'$. We need to show that any element of $(A \cap B)'$ is an element of $A' \cup B'$, and vice versa. But

$$\begin{aligned} x \in (A \cap B)' &\iff x \notin A \cap B \iff (x \in A \cap B)' \\ &\iff (x \in A \wedge x \in B)' \iff (x \in A)' \vee (x \in B)' \\ &\iff x \notin A \vee x \notin B \iff x \in A' \vee x \in B' \\ &\iff x \in A' \cup B', \end{aligned}$$

as required. With the exception of the fourth equivalence, every step here simply used the definitions of the various operations (intersection, union, complement, logical and, logical or, negation); the fourth equivalence used the the first of DeMorgan's laws from Table 3.1, namely, that $(p \wedge q)' \equiv p' \vee q'$. □

3.1.7 Duality (∗)

You may have noticed that some of the identities in Table 3.1 appear in pairs: there are two versions of the commutative, associative, distributive, and DeMorgan laws. This is certainly a handy mnemonic device, but it's more than that—it's based on a labor-saving device, called the "duality principle."[7]

Suppose that P is a proposition that only uses the operations $'$, $\wedge$, and $\vee$. If we replace all instances of $\wedge$, $\vee$, T, and F in p by $\vee$, $\wedge$, F, and T, respectively, we get a new proposition P^*, which is called the *dual* of P. For example:

$$\begin{aligned} \big(p \wedge (q \vee r)\big)^* &= p \vee (q \wedge r) \\ \big((p \wedge q) \vee (p \wedge r)\big)^* &= (p \vee q) \wedge (p \vee r). \end{aligned} \tag{3.4}$$

We then have the following important

Fact (First Duality Principle). *If two propositions (which only use the operations $'$, $\wedge$, and $\vee$) are equivalent, then their duals are equivalent.* □

Although a full proof is beyond the scope of this text, think of the first duality principle as being an extended version of De Morgan's laws.

Now suppose that P and Q are propositions with duals P^* and Q^*, respectively. We say that the propositional equivalence $P^* \equiv Q^*$ is the *dual* of the propositional equivalence $P \equiv Q$. Of course, it might be the case that neither of these equivalences is true.

Using the first duality principle, we now have the following

Fact (Second Duality Principle). *The propositional equivalence $P \equiv Q$ is true if and only if its dual equivalence $P^* \equiv Q^*$ is true.*

The second duality principle tells us that once we've established a given propositional equivalence, its dual equivalence is automatically true. Roughly speaking, we can use this principle to cut the amount of work we need to do in half.

Example 3.10: Let's see how we can save some work.

(a) On page 94, we proved the first distributive law

$$p \wedge (q \vee r) \equiv (p \wedge q) \vee (p \wedge r),$$

[7] Larry Wall, the inventor of the Perl programming language, is famous for his many aphorisms. One of my favorites is that a great programmer has three virtues; laziness, impatience, and hubris. Don't read this statement at the surface level; this is a special kind of laziness, impatience, and hubris!

which says that AND is distributive over OR. The second duality principle tells us that

$$(p \wedge (q \vee r))^* \equiv ((p \wedge q) \vee (p \wedge r))^*.$$

Using (3.4) and the equivalence above, we immediately see that the other distributive law

$$p \vee (q \wedge r) \equiv (p \vee q) \wedge (p \vee r),$$

is also true, i.e., that OR is distributive over AND.

(b) We have already proved one of De Morgan's laws, namely,

$$(p \vee q)' \equiv p' \wedge q'.$$

The second duality principle tells us that

$$((p \vee q)')^* \equiv (p' \wedge q')^*.$$

Since

$$((p \vee q)')^* = (p \wedge q)' \qquad \text{and} \qquad (p' \wedge q')^* = p' \vee q',$$

we see that

$$(p \wedge q)' \equiv p' \vee q',$$

i.e., the other version of De Morgan's law is true.

(c) Since the dual of

$$p \vee p' \equiv T$$

is

$$p \wedge p' \equiv F,$$

we would only need to prove one of these identities; the other one would follow immediately from duality. □

At first glance, duality would seem to be useful only for propositional equivalences involving complement ($'$), conjunction ($\wedge$) and disjunction ($\vee$). What about propositions involving *all* our connectives? The Boolean operations we need to deal with are exclusive or ($\oplus$), conditional ($\Rightarrow$) and biconditional ($\iff$)? These cases can be handled by first reducing them to complement, conjunction, and disjunction, via the rules

$$P \oplus Q \equiv (P \wedge Q)' \wedge (P \vee Q),$$
$$P \Rightarrow Q \equiv P' \vee Q,$$
$$P \iff Q \equiv (P' \vee Q) \wedge (P \vee Q').$$

(The first rule is from Exercise 3.3.12, while the second and third are the implication and biconditional laws found in Table 3.1.) After applying these rules, everything has been reduced to complement, conjunction, and disjunction, and so the duality principle may be applied.

Example 3.11: Determine the duals of $p \Rightarrow q$ and of $q' \Rightarrow p'$.

Solution: We find that

$$(p \Rightarrow q)^* \equiv (p' \vee q)' = p' \wedge q$$

and

$$(q' \Rightarrow p')^* \equiv \left((q')' \vee p'\right)^* \equiv (q \vee p')^* = q \wedge p'. \qquad \square$$

3.1.8 Indirect Proofs (∗)

> [W]hen you have eliminated the impossible, whatever remains, however improbable, must be the truth?
>
> Sherlock Holmes
> (Doyle, *Sign of the Four*)

Sometimes a direct, head-on approach to proving a statement works perfectly well. The following example shows such a "frontal attack" in action:

Example 3.12: Show that the square of an odd number is also an odd number.

Solution: Suppose that we let m denote our odd number; then we are being asked to show that m^2 is an odd number. Clearly, any odd number is "one off" from an even number, and an even number is the double of some integer. This means that we can write $m = 2n + 1$ for some other integer n. Squaring this equation and doing some algebra, we find that

$$m^2 = (2n + 1)^2 = 4n^2 + 4n + 1 = 2(2n^2 + 2n) + 1.$$

Let $k = 2n^2 + 2n$. Since $n \in \mathbb{Z}$, we know that $k = 2n^2 + 2n \in \mathbb{Z}$. This means that $m^2 = 2k + 1$ for some $k \in \mathbb{Z}$, which means that m^2 is an odd number. □

On the other hand, there are times when a frontal assault appears to be fruitless. When this happens, an indirect attack may be necessary. The simplest such technique is called *proof by contradiction*. Rather than directly proving that a statement is true, we assume (momentarily) that it is false, and see whether this assumption will lead to some kind of contradiction. Such a contradiction would mean that something we know to be true should be false, or vice versa.

Example 3.13: Show that the set $\mathbb{Z}$ of integers is infinite.

Solution: Suppose otherwise, i.e., that $\mathbb{Z}$ has finitely many elements. Under this assumption, it is clear that $\mathbb{Z}$ would have a biggest element, which we may call m. Since adding 1 to an integer always results in a new integer, we see that $m + 1$ is also an integer. But $m + 1 > m$, which contradicts the fact that m is the largest integer. Thus our assumption that $\mathbb{Z}$ is finite is false, which means that $\mathbb{Z}$ is infinite. □

Our second technique for indirect proofs is called a *contrapositive proof*. The *contrapositive* of an implication $p \Rightarrow q$ is the implication $q' \Rightarrow p'$. Recalling the contrapositive law

$$p \Rightarrow q \equiv q' \Rightarrow p'$$

from Table 3.1, we see that we can prove an implication by proving its contrapositive.

Example 3.14: Show that if the square of an integer is even, then the integer itself is even.

Solution: We are trying to show that

$$m^2 \text{ is even} \Rightarrow m \text{ is even} \qquad \text{for any } m \in \mathbb{Z}. \tag{3.5}$$

Feel free to try a direct proof of (3.5). Once you've tired of your efforts, you'll be willing to try a contrapositive proof. The contrapositive of (3.5) is

$$m \text{ is odd} \Rightarrow m^2 \text{ is odd} \qquad \text{for any } m \in \mathbb{Z}. \tag{3.6}$$

It just so happens that we've already shown that (3.6) is true in Example 3.12. So the contrapositive rule tells us that Example 3.14 is true, as well. □

Warning: The *converse* of the proposition $p \Rightarrow q$ is the proposition $q \Rightarrow p$. Note that the contrapositive of $q \Rightarrow p$ is $p' \Rightarrow q'$. The contrapositive law now tells us that the converse of $p \Rightarrow q$ is logically equivalent to $p' \Rightarrow q'$.

Do not confuse the concepts of "contrapositive" and "converse." Whereas the contrapositive of a proposition is logically equivalent to said proposition, the same is generally *not* true of the converse! (In other words, although there is a contrapositive law, there's no "converse law.")

For example, consider the proposition

any exact integer multiple of four is an even number,

whose converse is

any even number is an exact integer multiple of four.

The original statement is true; its converse is false. □

Bowing to tradition, let's close this subsection with

Example 3.15: Show that $\sqrt{2}$ is an irrational number.

Solution: Let's do a proof by contradiction. Suppose otherwise, i.e., that $\sqrt{2} \in \mathbb{Q}$. Then we can write $\sqrt{2} = p/q$ as a fraction. In this fraction, we can assume that p and q are positive integers, and that p and q cannot both be even numbers (if they were both even, we could reduce the fraction p/q to lower terms by dividing both the numerator p and denominator q by 2). We thus have

$$\begin{aligned}
\sqrt{2} = \frac{p}{q} &\Rightarrow \frac{p^2}{q^2} = 2 \\
&\Rightarrow p^2 = 2q^2 \\
&\Rightarrow p^2 \text{ is even} \\
&\Rightarrow p \text{ is even (by Example 3.14)} \\
&\Rightarrow p = 2r \text{ for some positive integer } r \\
&\Rightarrow (2r)^2 = p^2 = 2q^2 \qquad \text{(Remember that } p = 2q^2\text{!)} \\
&\Rightarrow 4r^2 = 2q^2 \\
&\Rightarrow 2r^2 = q^2 \\
&\Rightarrow q^2 \text{ is even} \\
&\Rightarrow q \text{ is even (again using Example 3.14)}
\end{aligned}$$

So both p and q are even, contradicting the fact that we chose p and q to not both be even. Since our assumption that $\sqrt{2} \in \mathbb{Q}$ leads to a contradiction, we must have $\sqrt{2} \notin \mathbb{Q}$, i.e., $\sqrt{2}$ is irrational. □

There is a legend of mathematical folklore involving this last example. You probably remember the Pythagorean theorem of geometry, which says that in a right triangle whose legs measure a and b units and whose hypotenuse measures c units, we always have $c^2 = a^2 + b^2$. In particular, this means that $\sqrt{2}$ is easily constructible as the hypotenuse of a right triangle, each of whose legs measures 1 unit. The Pythagoreans were philosophical followers of Pythagoras, dating back to the fifth century BCE, one of their tenets being that any constructible number must be a rational number, this idea being somehow related to the perfection of the integers. Apparently one of the Pythagoreans discovered the irrationality of $\sqrt{2}$, which disproved said foundational principle. Supposedly, his fellow Pythagoreans rewarded the person who discovered the irrationality of $\sqrt{2}$ (thereby invalidating the underpinnings of the Pythagorean philosophy) by pushing him to his death off the city wall.

3.1.9 From English to Propositions

Let's see how we can use propositional forms to capture the gist of logical arguments that appear in English. The main idea is to assign a Boolean variable to each phrase in the argument, connecting these variables with Boolean operations. Once we've gotten the hang of doing this, we can then use propositional laws, such as those in Table 3.1, to make logical deductions.

Example 3.16: I am a liberal and you are a conservative.

Solution: This one isn't too challenging. We let l denote the proposition "I am a liberal," and let c denote the proposition "You are a conservative." Then this statement can be restated as the propositional form $l \wedge c$. □

Example 3.17: If I am a liberal and you are a conservative, then he is a moderate.

Solution: This is a bit more complex. Let l, c, and m respectively denote the propositions "I am a liberal," "You are a conservative," and "He is a moderate." We can render this statement in the form $l \wedge c \;\Rightarrow\; m$; if you don't want to worry about the precedence rules, you could write this as $(l \wedge c) \;\Rightarrow\; m$. □

Example 3.18: I am a liberal and you are a conservative if and only if he is a moderate or she is a socialist.

Solution: This is even more complex. Let l, c, m, and s respectively denote the propositions "I am a liberal," "You are a conservative," "He is a moderate," and "She is a socialist." The propositional form $l \wedge c \iff m \vee s$ now captures our original statement; alternatively, you could write this as $(l \wedge c) \iff (m \vee s)$. □

Lewis Carroll is widely-known as being the author of the books *Alice's Adventures in Wonderland* and *Through the Looking Glass, and What Alice Found There*. You may not know that "Lewis Caroll" was the pseudonym of Rev. Charles Lutwidge Dodgson, who was a mathematician and logician. Carroll's book *Symbolic Logic* contains sixty *sorites* (logical puzzles), all of which show Caroll's sense of whimsy. Here's the first one in his book:

Example 3.19: Given the following facts:

(1) All babies are illogical.

(2) Nobody is despised who can manage a crocodile.

(3) Illogical persons are despised.

Prove that babies cannot manage crocodiles.

Solution: Let b, c, d, and l denote the status of being a baby, being able to manage a crocodile, being despised, and being logical. Our three facts may now be rewritten as

(1) $b \Rightarrow l'$.

(2) $c \Rightarrow d$.

(3) $l' \Rightarrow d$.

From (1) and (3), along with the transitive law, we have $b \Rightarrow d'$. Applying the contrapositive law to (2), we know that $d' \Rightarrow c'$. Since $b \Rightarrow d'$ and $d' \Rightarrow c'$, the transitive law tells us that $b \Rightarrow c'$. In other words, being a baby implies not being able to manage crocodile, as we were required to prove. □

Remark. The sorites of Example 3.19 has three facts. Carroll gradually adds more and more facts to the sorites found in his *Symbolic Logic*. The last sorites contains ten facts:

(1) The only animals in this house are cats;

(2) Every animal is suitable for a pet, that loves to gaze at the moon;

(3) When I detest an animal, I avoid it;

(4) No animals are carnivorous, unless they prowl at night;

(5) No cat fails to kill mice;

(6) No animals ever take to me, except what are in this house;

(7) Kangaroos are not suitable for pets;

(8) None but carnivora kill mice;

(9) I detest animals that do not take to me;

(10) Animals, that prowl at night, always love to gaze at the moon.

His conclusion? "I always avoid a kangaroo." □

3.1.10 Logic Circuits (∗)

We have already mentioned (page xiii) that without software, computers would be nothing but nice-looking doorstops. On the other hand, without hardware, we would be unable to run the software. So let's talk a little bit about computer hardware. How might one design the hardware in the first place? We shall see that Boolean logic plays a crucial role in answering this question.

Computers are built from a small list of basic electronic components, such as transistors, resistors, capacitors, and diodes. A fabrication technique called *very large scale integration* (VLSI), allows us to put a huge number of these components into a very small space, leading to the development of the computer chip as a self-contained component itself. These VLSI techniques have been improving at a rapid rate for a long time; as far back as 1965, Gordon E. Moore noted that the number of transistors that can placed on a computer chip tends to double every two years, a phenomenon that has since been given the name "Moore's law." Although some predict a near-term slowdown in the rate of growth predicted by Moore's law, the computer industry still continues to produce higher-performing microprocessors.

Since this area changes so rapidly, it would be foolish for us to link our discussion of computer hardware to a particular level of microprocessor technology.[8] After all, today's cutting edge may be tomorrow's obsolescence. Fortunately, there's no need to do so. We can build computer circuits out of simple *logic gates*, which are electronic components that evaluate the simplest basic logical operations. In such *logic circuits*, the

[8]Moreover, this is not a textbook about electrical engineering, and so such a discussion wouldn't be fitting in the first place.

absence of an electrical signal signifies a Boolean value "false," whereas the presence of an electrical signal signifies "true." We emphasize that we're not going to worry about how one constructs a particular kind of logic gate; we'll take their availability as a given.

Remark. Recall that on page 79, we said that we would use "0" and "1", rather than "F" and "T", to represent "false" and "true" when dealing with logic circuits. We now see that this choice can be motivated by the fact that absence or presence of electrical signal signifies "false" and "true" (respectively) in logic circuits. □

Recall that $\{', \wedge, \vee\}$ is a *universal* set of Boolean operations, i.e., we can create any Boolean expression using these operations (see page 86). Thus, let us suppose that the following kinds logic gates, whose diagrams may be found in Figure 3.1, are available to us:

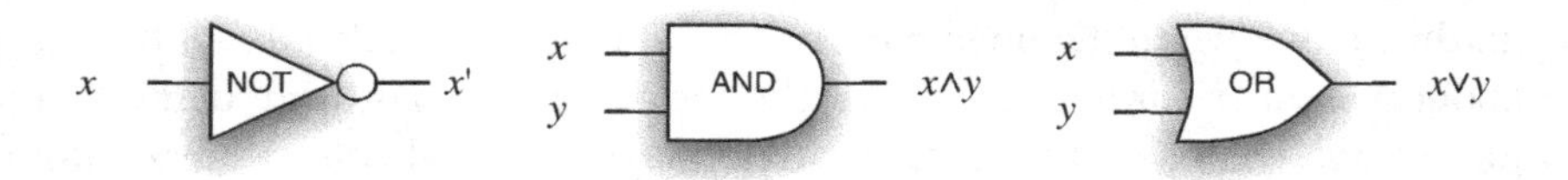

Figure 3.1: Basic logic gates: NOT, AND, OR

- The NOT-gate (also called an *inverter*). This gate has one input wire and one output wire. If the input wire has a given value x, then its output wire will have the logical negation x' of x.
- The AND-gate. This gate has two input wires and one output wire. If x and y are the values at the two input wires, the $x \wedge y$ will be the value at the output wire.
- The OR-gate. This gate has two input wires and one output wire. If x and y are the values at the two input wires, the $x \vee y$ will be the value at the output wire.

It now follows that we can build any logic circuit using NOT-, AND-, and OR-gates.

Example 3.20: How might we build a circuit that computes the proposition $x \vee y'$?

Solution: The solution requires one NOT-gate (for computing y' from the input value y), as well as one OR-gate, which takes the value of y' and the value of x as its inputs, thereby producing the value $x \vee y'$. The desired logic circuit is shown in Figure 3.2. □

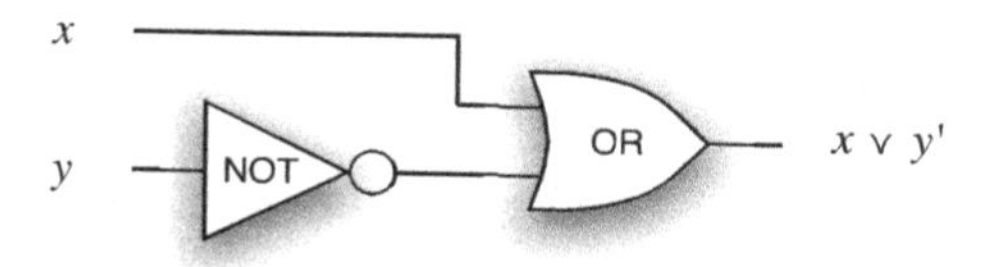

Figure 3.2: Logic circuit for computing $x \vee y'$

Example 3.21: What is the Boolean expression corresponding to the logic circuit given in Figure 3.3?

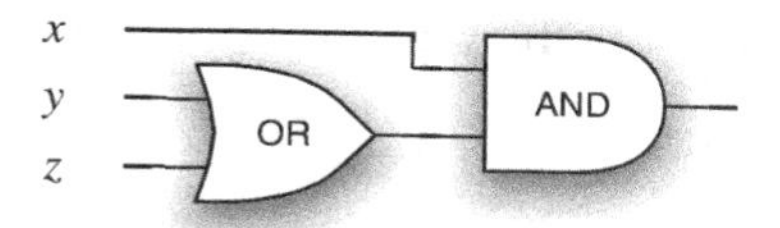

Figure 3.3: Logic circuit for Example 3.21

Solution: This circuit computes the Boolean expression $x \wedge (y \vee z)$. □

Example 3.22: How would we build an XOR-circuit?

Solution: From Exercise 3.3.12, we see that $x \oplus y$ is logically equivalent to

$$(x \wedge y)' \wedge (x \vee y).$$

The desired logic circuit is shown in Figure 3.4. □

Example 3.23: A building is protected by a fire alarm, a smoke alarm, and a burglar alarm. If either the fire alarm or the smoke alarm is set off, then the fire department is called. Moreover, if any of these three alarms is set off, then the police is called. Design a logic circuit that controls this alarm system. Don't worry about the mechanics of setting the various alarms or calling the fire department or the police; simply deal with the logic of which department gets called, under which circumstances.

Solution: This alarm system is controlled by three inputs:

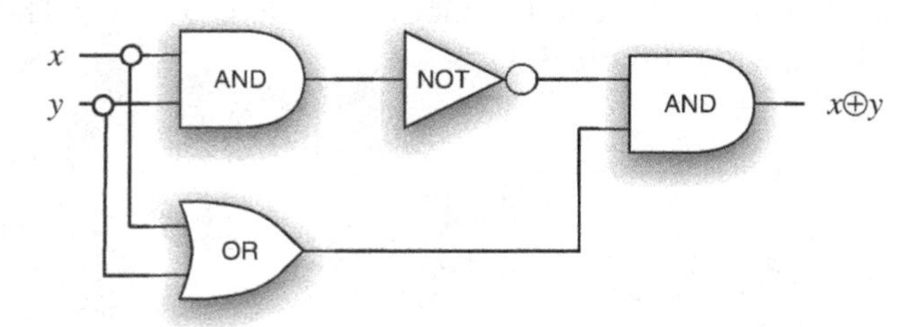

Figure 3.4: Logic circuit for computing exclusive-or

- the signal coming from the burglar alarm, which we will denote by the variable b,
- the signal coming from the fire alarm, denoted f, and
- the signal coming from the smoke alarm, denoted s.

There will be two outputs:

- a signal telling us to call the fire department,[9] which we'll denote by a, and
- a signal telling us to call the police department, which we'll denote by p.

Thinking about this a bit, it should be clear that

$$p = b \vee f \vee s \qquad \text{and} \qquad a = f \vee s.$$

The desired logic circuit is shown in Figure 3.5. □

3.2 Predicate Logic (∗)

So far, we have been working with propositions—statements that are either true or false. Sometimes, we are faced with what (for lack of a better term) we might consider "propositions containing variables." Let's illustrate this point by using a famous classical syllogism:

[9] We can't use f to represent this signal, since it's already in use. The letter a is the first letter of "aish," the Hebrew word for "fire." The first letter of the word for "fire" in various other languages (French, German, Greek, Russian) was either "f" or "p," so I couldn't use them. So it goes.

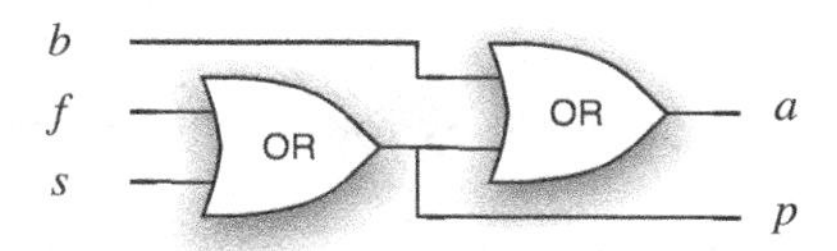

Figure 3.5: Logic circuit for an alarm system

- All men are mortal.
- Socrates is a man.
- Therefore Socrates is mortal.

If we wanted to use symbolic logic to capture this particular argument, we might define

$$\text{man}(x) = \text{“}x \text{ is a man,”}$$

and

$$\text{mortal}(x) = \text{“}x \text{ is mortal.”}$$

Since Socrates is (was?) a man, we'd agree that

$$\text{man(Socrates)}, \tag{3.7}$$

which is a concise way of saying "man(Socrates) is true." Presumably, we'd all agree that

$$\text{man}(x) \Rightarrow \text{mortal}(x) \text{ for any person } x. \tag{3.8}$$

From these two statements, one would naturally conclude

$$\text{mortal(Socrates)}, \tag{3.9}$$

and so Socrates is mortal.

There are a couple of lessons we can draw from this. The first is that propositions with variables, such as man(x) and mortal(x), are useful. The second is that the reasoning that derives the conclusion (3.9) from the hypotheses (3.7) and (3.8) is completely general, and has nothing to do with Socrates, humanity, or mortality.

So let's try to tighten this all up. Let us say that a *predicate* is an expression containing one or more variables that becomes a proposition whenever we substitute specific values for the variables. In other words, a predicate is almost like a proposition, except that it has one or more variables; when we make any appropriate substitution for these variables, we always get a proposition.

Example 3.24: Let

$$\text{four}(x) = \text{“}x \text{ is divisible by four.”} \qquad \text{for any } x \in \mathbb{Z}.$$

Solution: If we replace the variable x by any particular integer, we'll get a proposition. The table

x	four(x)	truth value of four(x)
⋮	⋮	⋮
-4	-4 is divisible by 4	T
-3	-3 is divisible by 4	F
-2	-2 is divisible by 4	F
-1	-1 is divisible by 4	F
0	0 is divisible by 4	T
1	1 is divisible by 4	F
2	2 is divisible by 4	F
3	3 is divisible by 4	F
4	4 is divisible by 4	T
⋮	⋮	⋮

helps us to envision this process. Although this table gives truth values associated with a variable, this is not a truth table. The variable here is not a Boolean variable, but an integer variable, which means that it can take any integer value. The middle column gives the value of the predicate four(x) for any integer value x, which means that this column is a list of propositions. Each of these propositions has a truth value, as seen in the final column. □

3.2.1 Quantifiers

Before we start talking about how to reason using predicates, we'll need some way to give them truth values. This is a little self-contradictory, since a predicate does not have a truth value! However, if we could turn a predicate into a proposition, then we would be able to assign a truth value to that proposition. This transformation of predicates into propositions can be done by a process called *quantification.*

Suppose that $p(x)$ is a predicate in the variable x, where x can take values in a set S. There are two forms of quantification:

Universal quantification: Here, we ask that the predicate $p(x)$ be true for *all* values of $x \in S$. We use $\forall$ (an upside-down A, which reminds us of the word "all") as the symbol for universal quantification. So we let

$$\forall x \in S,\ p(x)$$

denote the proposition "For all elements $x \in S$, $p(x)$ is true."

Existential quantification: Here, we ask that the predicate $p(x)$ be true for *some* value of $x \in S$; it might turn out that there's only *one* such value. We use $\exists$ (an upside-down E, which reminds us of the word "exists") as the symbol for existential quantification. So we let

$$\exists x \in S\colon\ p(x)$$

denote the proposition "There exists some element $x \in S$ such that $p(x)$ is true."

Warning: Notice that the punctuation for universal quantification is not the same as that for existential quantification. We use a comma in

$$\forall x \in S,\ p(x)$$

because we use a comma in the analogous English statement. We use a colon in

$$\exists x \in S\colon\ p(x)$$

as an abbreviation for the phrase "such that", which appears in the analogous English statement.[10] You will sometimes see other notations, such as

$$\forall x \in S\,\big(p(x)\big)$$

and

$$\exists x \in S\,\big(p(x)\big)$$

that don't use the comma or colon. □

Example 3.25: As in Example 3.24, let

$$\text{four}(x) = \text{“}x \text{ is divisible by four.”} \qquad \text{for any } x \in \mathbb{Z}.$$

The proposition

$$\forall x \in \mathbb{Z},\ \text{four}(x)$$

[10]Recall that we used a colon for the same purpose when describing set-builder notation, see page 8.

is false, since not all integers are divisible by 4 (look at the table). On the other hand, the proposition

$$\exists x \in \mathbb{Z}: \text{four}(x)$$

is true, since there exists an integer that is a multiple of 4 (again, look at the table). □

Example 3.26: [SN] Look at Table A.2. You'll see that every person having a postgraduate (master's or doctoral) degree also has a bachelor's degree. Letting Name denote the set of names in the table, we define bachelors(x) to mean "x has a bachelor's degree" and postgrad(x) to mean "x has a postgraduate degree." Then

$$\forall x \in \text{Name}, \text{postgrad}(x) \implies \text{bachelors}(x).$$

This shouldn't surprise you very much. In the normal course of events, people earn their bachelor's degree before they go on to a postgraduate degree (after all, they *are* called "postgraduate degrees"). If we let P denote the set of all people, we would conjecture that

$$\forall x \in P, \text{postgrad}(x) \overset{?}{\implies} \text{bachelors}(x).$$

However, this conjecture is false. Some universities will allow a prospective graduate student to substitute life experience or some other form of learning for a bachelor's degree, and so you'll occasionally find somebody with (say) a PhD degree without having earned a bachelor's degree. □

3.2.2 Some Rules for Using Predicates

Now suppose that $p(x)$ and $q(x)$ are predicates involving the variable x, where x can take values in some set S. Suppose further that

$$p(x) \Rightarrow q(x) \qquad \text{for any } x \in S,$$

in other words, that the truth of $p(x)$ (for any particular value of $x \in S$) implies the truth of $q(x)$. Finally suppose that we know that $p(a)$ is true for some particular $a \in S$. It then follows that $q(a)$ is also true. We can summarize this by saying

$$[(\forall x \in S, p(x) \Rightarrow q(x)) \wedge p(a)] \Rightarrow q(a). \tag{3.10}$$

This is the *classical syllogism*, which we used at the beginning of this section to derive (3.9) from (3.7) and (3.8).

We are sometimes faced with the negation of a quantified statement. (For example, we might want to say that a particular classical syllogism is false.) Such statements can be simplified if we apply the *negation laws*

$$[\exists x \in S\colon\ p(x)]' \equiv [\forall x \in S,\ p'(x)]$$

and

$$[\forall x \in S,\ p(x)]' \equiv [\exists x \in S\colon\ p'(x)].$$

Let's see why the first of these is true; the second one is analogous. The statement "$[\exists x \in S\colon\ p(x)]'$" means that it is false that $p(x)$ is true for some $x \in S$. In other words, this statement says that $p(x)$ must fail to be true for *any* $x \in S$, which is another way of saying "$\forall x \in S,\ p'(x)$."

Of course, predicates can have more than one variable.

Example 3.27: On New Year's Day, you may see a photograph of the Polar Bears' Club in the newspaper. This is a group of hardy souls who think that it's a wonderful idea to go swimming in the Atlantic Ocean on one of the coldest days of the year. Let's see how we can put quantification to work in discussing the exploits of the Polar Bears' Club.

Let P be a set of people, and let T be a set of temperatures. (For example, P might be the set of people living in Brooklyn, NY and T might be the set[11] $\{-4, -3, -2, -1, 0, 1, \ldots, 104\}$.) We might let "beach$(p, t)$" mean that person p will go to the beach if the temperature reaches t degrees. Since we have two variables (p and t), we have the following quantification choices:

- beach(p, t), i.e., no quantification at all. This is a two-variable predicate.
- We can quantify in one variable. Quantifying over p gives the following predicates in t:
  - $\exists p \in P\colon$ beach(p, t), i.e., "some person p will go to the beach at temperature t."
  - $\forall p \in P$, beach(p, t), i.e., "when the temperature reaches t, everybody will go to the beach."

[11]This reflects the lowest and highest recorded temperatures for Brooklyn, NY as of the date this was written. (Source: The Weather Channel website.)

Quantifying over t gives the following predicates in p:

- $\exists t \in T$: beach(p, t), i.e., "there's some temperature at which person p will go to the beach."
- $\forall t \in T$, beach(p, t), i.e., "person p will go to the beach, no matter what the temperature may be."

Note that there's a certain amount of variability in how we might state these predicates in English.

- We can quantify in both variables, getting eight different propositions; we'll let you figure out what these all mean (some of these are part of Exercise 3.3.19).

$$\exists p \in P\colon [\exists t \in T\colon \text{beach}(p, t)]$$
$$\exists p \in P\colon [\forall t \in T, \text{beach}(p, t)]$$
$$\forall p \in P, [\exists t \in T\colon \text{beach}(p, t)]$$
$$\forall p \in P, [\forall t \in T, \text{beach}(p, t)]$$
$$\exists t \in T\colon [\forall p \in P, \text{beach}(p, t)]$$
$$\forall t \in T, [\exists p \in P\colon \text{beach}(p, t)].$$

Note that for "mixed" quantification (containing both $\forall$ and $\exists$, the order matters. By the way, most people would omit the brackets. □

Example 3.28: As a counterweight to the Polar Bears' Club, we might think about the people who perhaps spend far too much time online than is healthy. We'll re-use some of the notation in Example 3.27, letting P be a set of people, and let T be a set of temperatures. We now might let "online(p, t)" mean that person p will be online (rather than going to the beach) if the temperature reaches t degrees. We would then have the following quantification choices, whose restatements in English we'll let you figure out on your own:

- online(p, t), i.e., no quantification at all. This is a two-variable predicate.
- We can quantify in one variable. Quantifying over p gives the following predicates in t:

$$\exists p \in P\colon \text{online}(p, t)$$
$$\forall p \in P, \text{online}(p, t).$$

Quantifying over t gives the following predicates in p:

$$\exists t \in T: \text{ online}(p, t)$$
$$\forall t \in T, \text{online}(p, t).$$

- We can quantify in both variables, getting the propositions:

$$\exists p \in P: \exists t \in T: \text{ online}(p, t)$$
$$\exists p \in P: \forall t \in T, \text{online}(p, t)$$
$$\forall p \in P, \exists t \in T: \text{ online}(p, t)$$
$$\forall p \in P, \forall t \in T, \text{online}(p, t)$$
$$\exists t \in T: \forall p \in P, \text{online}(p, t)$$
$$\forall t \in T, \exists p \in P: \text{ online}(p, t).$$

(Note that we didn't bother with the brackets.) □

3.3 Exercises

3.3.1 Give the truth table of $(p')'$.

3.3.2 Give the truth table of $p \Rightarrow q'$.

3.3.3 Use a truth table to show that $(p \Rightarrow q')' \vee (q \Rightarrow p')$ is a tautology.

3.3.4 (∗) Draw the parse tree of the expression

$$p \vee (q \wedge r) \iff (p \wedge q) \vee (p \wedge r).$$

3.3.5 Suppose that l, m, and w respectively denote the propositions "Linux is good," "Mac OS is good", and "Windows is good."

(a) Translate "Windows is evil, Linux is good, and Mac OS is good." into a propositional form.

(b) Translate "Windows is evil. Moreover, Linux is good or Mac OS is good." into a propositional form.

(c) Prove that the statements "Windows is evil. Moreover, Linux is good or Mac OS is good." and "Windows is evil and Linux is good, or Windows is evil and Mac OS is good." are equivalent.

3.3.6 Let

- b represent "John is a business major"
- c represent "John is a computer science major"
- m represent "John is a math major"
- r represent "John will be rich"
- s represent "John is smart"

Write each of the following as propositions, using only the five variables b, c, m, r, and s.

(a) John is a computer science major and a math major.

(b) John is a math major and is not rich.

(c) If John is a smart business major, then John will be rich.

(d) If John is a computer science major or a math major, and he is not a business major, then John is smart.

(e) If John is business major who is not smart, then John will not be rich.

3.3.7 $(*)$ What is the contrapositive of the following statements?

(a) If Caesar is power-hungry, then Brutus will be among his assassins.

(b) If my mother-in-law is cold, she'll tell my wife to put on a sweater.

(c) If my kilt has a rounded corner, then it will not need a fringe.

(d) If poultry could sing, then my cat would enjoy opera.

(e) If cars could run on broccoli, then international politics would be different.

3.3.8 $(*)$ Find the converse of all the statements in Exercise 3.3.7.

3.3.9 Use a truth table to prove the following:

(a) The first of DeMorgan's laws: $(p \wedge q)' \equiv (p') \vee (q')$.

(b) The second distributive law: $p \vee (q \wedge r) \equiv (p \vee q) \wedge (p \vee r)$.

(c) The implication law: $(p \Rightarrow q) \equiv (p' \vee q)$.

3.3.10 Show that $\Rightarrow$ is *not* associative, i.e., that

$$[(p \Rightarrow q) \Rightarrow r] \not\equiv [p \Rightarrow (q \Rightarrow r)].$$

3.3.11 The NAND ("not and") operator $\barwedge$ is defined as

p	q	$p \barwedge q$
T	T	F
T	F	T
F	T	T
F	F	T

Note that

$$p \barwedge q \equiv (p \wedge q)'.$$

Try to do the following *without* resorting to truth tables.

(a) Prove that $p' \equiv p \barwedge p$.

(b) Prove that $p \wedge q \equiv [(p \barwedge q) \barwedge (p \barwedge q)]$.

(c) Show that $p \vee q$ can be expressed in terms of the NAND operation.

Hint: Reduce $p \vee q$ to a proposition that only uses $'$ and $\wedge$.

Congratulations! You've just proved that $\{\barwedge\}$ is a one-element universal set!

3.3.12 Prove that

$$(p \oplus q) \equiv [(p \wedge q)' \wedge (p \vee q)].$$

3.3.13 ($*$) Find the duals of the following propositions.

(a) $p \wedge p'$.

(b) $p \wedge (q \vee r)$.

(c) $(p \wedge q') \vee (p' \wedge r)$.

(d) p'.

3.3.14 ($*$) Use duality to provide a *very* simple proof of the contrapositive law

$$(p \Rightarrow q) \equiv (q' \Rightarrow p')$$

Hint: See Example 3.11.

3.3.15 ($*$) Suppose that the average of ten numbers is 42. Use proof by contradiction to show that at least one of these numbers is greater than or equal to 42.

3.3.16 ($*$) Show that if an integer m^3 is even, then m is even.

3.3.17 Given the following facts:

(1) No ducks waltz.

(2) No officers ever decline to waltz.

(3) All my poultry are ducks.

Prove that my poultry are not officers. (This is sorites # 5 in Carroll's *Symbolic Logic*.)

3.3.18 ($*$)A fire alarm contains two smoke detectors. If either of the smoke detectors is activated, a sprinkler system is turned on. If both of the smoke detectors are activated, then fire department is called. Design a logic circuit for this fire alarm.

3.3.19 ($*$) Write out each of these predicates in English (see Example 3.27). Try to make your English as idiomatic as possible; for example, try to avoid using the phrase "there exists."

(a) $\exists\, p \in P\colon [\exists\, t \in T\colon \text{beach}(p, t)]$.

(b) $\exists\, p \in P\colon [\forall\, t \in T, \text{beach}(p, t)]$.

(c) $\forall\, p \in P, [\exists\, t \in T\colon \text{beach}(p, t)]$.

(d) $\forall\, p \in P, [\forall\, t \in T, \text{beach}(p, t)]$.

3.3.20 ($*$) **[SN]** Let employed(x) mean that x is employed. Using the data in Tables A.2 and A.3, determine whether the statement

$$\forall x \in \text{Name}, \text{bachelors}(x) \implies \text{employed}(x)$$

is true or false.

3.3.21 ($*$) Suppose that $g(x)$ denotes the predicate "x is good." Furthermore, suppose that good and evil are opposites.

(a) Translate "Windows is evil. Moreover, Linux is good or Mac OS is good." into a propositional form.

(b) Translate "Windows is evil and Linux is good, or Windows is evil and Mac OS is good." into a propositional form.

3.3.22 ($*$) Abraham Lincoln said:

> You can fool all of the people some of the time, and you can fool some of the people all of the time, but you can't fool all of the people all of the time.

(a) Express this statement using the notation of predicate logic.

(b) Express the negation of this statement in colloquial English.

Hint: Believe it or not, the simplest way to do this is to use the negation rules in Section 3.2.2 to determine the logical negation of your answer to (a), and to then translate this result into English.

Chapter 4

Relations

> When we try to pick anything out by itself, we find it hitched to everything else in the universe.
>
> John Muir

Sets give us a rigorous way to talk and reason about collections of objects. Logic gives us a rigorous way to talk about conditions and decisions. Now we will discuss how to describe the way objects can relate to each other in a formal manner.

The concept of relations and the material in this and the subsequent chapter on *functions* are indispensable concepts in building social networking sites. When we think of a friend or family member, a flood of connections might gush through our mind: their appearance, their likes, their location, their travels and so forth. Each connection, trickling through the synapses of our brain, finds other, like connections. You might, on thinking of your sister, recall that she has red hair. Thinking of this, you might become focused on what color hair the other members of your family have—is your sister the only red-head? On a social networking site, our interests, preferences, location and so forth are linked to our profile and these characteristics can be shared by our friends and acquaintances. To provide the ability to represent and reason about these interrelated characteristics we have to introduce a more formal framework for *relating* one set of things to another: the concept of *relations*.

In fact, the use of relations as a way to represent structured knowledge pre-dates social networks. It is very likely that you have heard the word *database* before, and given the pervasiveness of computing devices, it's also likely that you have a good, if informal idea of what a database is. A collection of information stored on a computer is often naively called a database. To understand why that common misconception is not a sufficiently useful definition of a database, consider the following collection:

{Alex, Topeka, Hartford, Alyssa, F,
Albany, S, 02/15/1996, M, 02/01/1964}

This is certainly a collection of information (and recall that we consider a set to be a formal way to describe collections). However, there is no way to understand how the various pieces of information in the collection are related to one another, and therefore, no way to use this collection as a resource to answer a question (other than to inspect all the elements one by one, of course). Now, take a look at the first two lines in Table A.1 of the Appendix. That table contains the same information, but in a much more useful form! If you need to answer a question about Alex or Alyssa's hometown or birthday, the way the information is organized into a table makes this easy! Now you can see why a database is more correctly defined as a collection of information *structured* for access (e.g., answering a question), management and update (easily adding more information). The extra structure in Table A.1 (the table columns and table headings) tells you how the pieces of information are related to each other.

When we talk about relations, we will introduce several useful types of relations. Knowing whether a relation falls into one of these types allows us to construct more efficient computer implementations: For example, if we know that the sibling relation (e.g., Pete is a sibling of Mary) is *reflexive* or *symmetric* we can spend less money on computer equipment by not storing redundant information. If we know that a relation is *transitive* this allows us to "discover" new members of the relation for free! In fact this idea lies at the core of how family connections are discovered, or new potential friends recommended, on some social networking sites.

This chapter will leverage the tools already developed in the chapters on sets and logic to show how relations can be represented and used for representing knowledge and solving problems.

4.1 Ways to Describe Relations Between Sets

A *relation* is a connection between elements in one set and elements in a second set. Consider the following example of a relation that connects the names of people to the ages of those people. Let the first set be the set of names of people in your class, and the second set be the set of natural numbers. The connection in this relation links the name to the age of the person named. We might pick an intuitive name for this relation, e.g., *age*, since the relation allows us to "connect" each person's name to their age. Note that it can make a difference which set goes first! For this reason we will give the two sets different names: we will call the first set the *domain* of the relation and the second set the *codomain* of the relation. In some interesting and useful cases the domain and codomain are the same set. In that case we will simply refer to the relation as being "on" the set, rather than spelling out the domain and codomain.

There are several ways we can choose to describe a relation like *age*.

- We can describe it in English as we did in the previous section. This has the advantage of being easy to understand, but the disadvantage of being open to misinterpretation.

- We can also describe it as a picture, by drawing the *domain* and *codomain* sets with arrows between the connected elements in each. This also has the advantage of being easy to grasp, but the disadvantage of being limited to what you can fit on a page! This is a problem for infinite sets such as $\mathbb{N}$.

- Another visual way to describe a relation is as a *table* of values (similar in concept to a train schedule). All the connected pairs are shown in adjacent columns (or rows) in the table. This has similar advantages and disadvantages to drawing a picture, but is certainly a very familiar and easy to understand representation, e.g., Table A.1.

- Finally, we can represent a relation using the Cartesian product and set builder notation. This has the advantage of supporting a rigorous definition of a relation, but the disadvantage of being the least easy representation to grasp at first sight.

We will look at all of these representations and determine how to switch between them. When we have to solve a problem using relations, we will pick the representation that seems most natural to us.

4.1.1 Using English

To define a relation in English we need to specify three things:

- the *domain* of the relation (in language terms, the "subject" of the relation),

- the *codomain* of the relation (in language terms, the "object" of the relation), and
- the connection or *rule* that links the elements in the domain to elements in the codomain.

We will refer to this ordered pair, one element of the domain linked to one element of the codomain, as an *element* of the relation. We will write it using the notation that we have introduced previously for ordered pairs, e.g., (x, y) where x is the element of the domain and y the element of the codomain.

Example 4.1: What elements are in the following relation?

Domain:	the set of names of people in your family
Codomain:	{red, black, brown, blond, flaxen, pink, green}
Rule:	(x, y) is in the relation if and only if x's hair color is y.

Solution: To complete this example, you need to have information about the people in your family. Let's imagine that you have a family of five: Mary (Mom), Pete (Dad), Molly, Sandra and Mark. All are brown haired except for Molly, whose hair is red. In that case, we can write our answer:

Domain ={Mary, Pete, Molly, Sandra, Mark}
Codomain ={red, black, brown, blond, flaxen, pink, green}
Elements are: (Mary, brown), (Pete, brown), (Molly, red)
(Sandra, brown), (Mark, brown)

You should take a moment and create this list for your family. That exercise will illustrate some important facts about relations:

- You will see that it's possible that two of your family members will have the same hair color, and hence the rule will connect two different people to the same element of the *codomain*.
- In many families, no member has green hair! In which case, no element of the domain will ever connect to green in the codomain. A relation may have elements of its codomain that are not connected to any element of its domain. □

Example 4.2: What elements are in the following relation?

Domain:	{Molly, Sandra, Mark}
Codomain:	{Molly, Sandra, Mark}
Rule:	(x, y) is in the relation if and only if x is the sister of y.

Solution: This relation's rule says that elements are linked if the first is the sister of the second. Of course, if you do not know this family information then you cannot tell which elements to link! Let us assume it is the same family that we have introduced in the previous example. In that case, we can write the answer as

Elements are: (Sandra, Molly), (Molly, Sandra), (Sandra, Mark),
(Molly, Mark)

Note that Mark can appear as a second element in a relationship pair, but not as the first element. □

In Example 4.1, the domain and codomain are different sets, whereas in Example 4.2, they are the same. There are many important examples in which the domain and the codomain are the same set; when this happens, we will economize things a bit, referring to the relation as a *relation on* the given set. So we could have simply described Example 4.2 as a relation on the set {Molly,Sandra, Mark} defined by the rule

(x, y) is in the relation if and only if x is the sister of y.

These first two examples illustrate well the fact that using English to specify relations is quite informal but easy to grasp. These relations correspond fairly closely to the use of the word "relation" (or "relationship") in English. However, we can also specify relations between numbers. Note that a relation between numbers is often described as a relation *on* the particular set of numbers, as in the next example:

Example 4.3: Define a relation on the set $\mathbb{N}$ of natural numbers by the rule

(x, y) is in the relation if x is one more than y (i.e., if $x = y + 1$).

What elements are in this relation?

Solution: For this example, since we understand what is in $\mathbb{N}$ and we know what addition is, we have all the information we need to say what is in the relation: $(1, 0)$ is in the relation, as is $(2, 1)$ and $(3, 2)$ and $\ldots$. The fact that this is an infinite list means that we can only give the first few examples of what is in the relation. We use an ellipsis to indicate the remaining elements. □

Example 4.4: Using the notation from the examples above, determine what is in the relation defined by

Domain:	people in your immediate family
Codomain:	$\mathbb{N}$ (the natural numbers)
Rule:	(x, y) is in the relation if "x is y years old"

Solution: This relation will be a list of ordered pairs. The first element of each ordered pair will be the name of a person in your family. The second element of that ordered pair will be the age of that person in your family. For example, if your family consists of three people: Mike, Susie and Jimmy, where Mike is 35 years old, Susie is 29 years old and Jimmy is 5 years old, then the relation would include only (Mike, 35), (Susie, 29), and (Jimmy, 5). □

Example 4.5: Using the notation from the examples above, determine what is in the relation on the natural numbers $\mathbb{N}$ defined by the rule

$$(x, y) \text{ is in the relation if } x \text{ is ten more than } y \text{ (i.e., if } x = y + 10).$$

Solution: The natural numbers is an infinite set, so our relation in this case will also be infinite. We will list the first few elements, and once we have listed enough that a reader can determine the pattern, then we will employ the ellipsis to indicate the remainder of the relation. The smallest natural number is 0. By the rule of this relation, 0 should be paired with 10, since 10 is ten more than 0. Therefore we would write $(10, 0)$ as our first element in the relation. The next biggest natural number is 1. By the rule of the relation, 1 should be paired with 11, since 11 is ten more than 1. Therefore we would write $(11, 1)$ as the second element of our relation. Following this reasoning for a few more numbers, we write the full list of elements in the relation as $(10, 0), (11, 1), (12, 2), (13, 3), \ldots$. □

4.1.2 Using a Picture

Sometimes it helps to view a relation as a picture: We can draw the solution to Example 4.2 as shown in Figure 4.1: The elements of the domain and codomain are shown as labeled circles, while connections between elements in the relation are indicated by joining those elements with an arrow. For each (x, y) in the relation, the arrow starts at x and ends at y.

Consider the relation shown in Figure 4.2. The relation contains (a, a) and this is indicated graphically by putting a "loop" at the element a, i.e., an arrow starting and ending at the a. When you show a relation graphically, there is no need to state the rule for the relation (as we did in the previous section) since the rule is implicit in the arrows that are drawn.

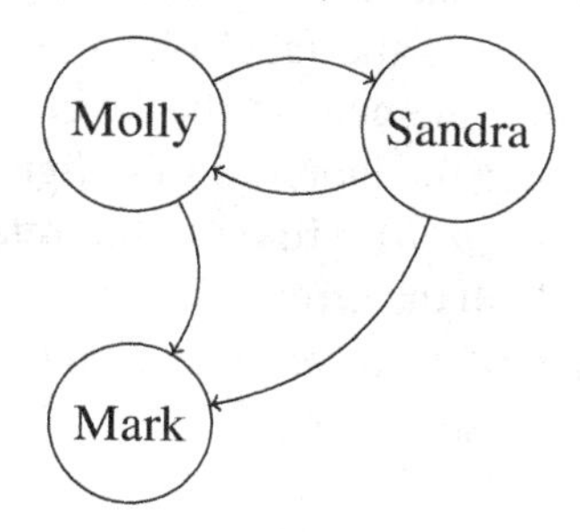

Figure 4.1: A graphical representation of the relation in Example 4.2.

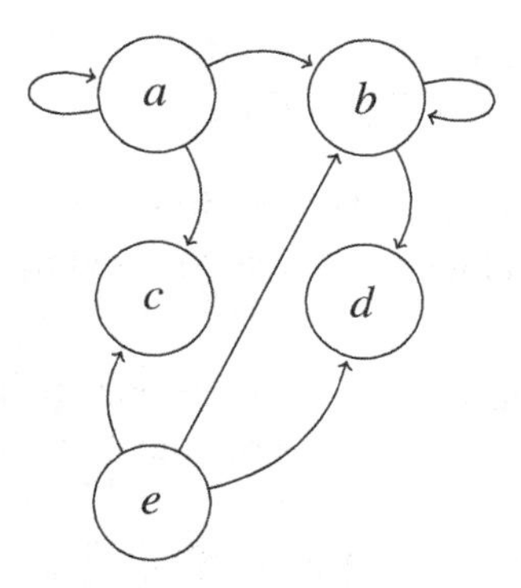

Figure 4.2: A graphical representation of a relation.

Example 4.6: Write out the domain, the codomain, and all the elements of the relation in Figure 4.2.

Solution: The domain and codomain are the same in this example, namely, the set $\{a, b, c, d, e\}$. The relation is a set of ordered pairs, where the first element of the ordered pair is the label of a circle from which an arrow originates, and the second element is the label of a circle at which the arrow terminates. Consider the label a in the figure. An arrow originates at a and goes to b, and so (a, b) will be in the relation, as will (a, a) and (a, c). That disposes with all the arrows that start on a. Now we move to b, and then c, and so forth until we have listed all the pairs in the relation. The result can be written as the set

$$\{(a, a), (a, b), (a, c), (b, b), (b, d), (e, b), (e, c), (e, d)\}. \qquad \square$$

4.1.3 Using a Table

Another visual method to describe a relation is a *table* similar to a phone book or train timetable. The elements of the domain of the relation are listed in the top row of the table. The elements of the codomain are listed in the bottom row of the table. Each column of the table corresponds to an ordered pair in the relation.

Example 4.7: Show the elements of the relation of Figure 4.2 as a table.

Solution:

a	a	a	b	b	e	e	e
a	b	c	b	d	b	c	d

□

Example 4.8: [SN] Look at the first four rows of Table A.1 in the Appendix. Write out the relation "lives in a city called" and the relation "is of the sex" as a table.

Solution: The information from columns one and two in Table A.1 provide all the information we need to write out the "lives in a city called" relation:

Alex	Alyssa	Angela	Anna
Topeka	Hartford	Charlotte	Hartford

The information from columns one and four in Table A.1 provide all the information we need to write out the "is of the sex" relation:

Alex	Alyssa	Angela	Anna
F	F	F	F

□

Example 4.9: [SN] Look at the first four rows of Table A.1 in the Appendix. How many relations are defined by those rows of the table?

Solution: A relation from one set to another set is called a *binary* relation. The elements of a binary relation will be ordered pairs. There are many binary relations defined by the first four rows of Table A.1. For example, the binary relation

{(Alex, Topeka), (Alyssa, Albany), (Angela, Denver), (Anna, Hartford)}

connects a person's name to the name of his or her hometown. Similarly there is a binary relation that connects a person's name to his or her birthdate, and another that connects a person's name to his or her marital status, and so forth. But notice there are also other relations specified by the table: There is a relation between cities and hometowns, an accidental relationship, but nonetheless it constitutes a relation (and we will come back to this situation in the next chapter when we discuss a special kind of relation called a function). In fact, each column specifies a binary relation with each other column in the table. Since there are 6 columns in the table, that means there are 6×6 possible binary relations!

There are other kinds of relations as well as binary relations. A *ternary* relation is a relation between three sets, and its elements are ordered triplets. In fact, a *k-ary* relation is a relation between k sets (e.g., a 6-ary relation or even a (gulp) 101-ary relation). There is one intuitive 6-ary relation specified by Table A.1:

{ (Alex, Topeka, Topeka, F, 02/15/1996, S)
(Alyssa, Hartford, Albany, F, 02/01/1964, M)
(Angela, Charlotte, Denver, F, 06/15/1967, F)
(Anna, Hartford, Hartford, F, 5/19/1989, U) }

Does this look familiar? The relation when represented as a set of 6-tuples looks almost identical to the original table! The concept of the *relational database* arises from this observation. It is impossible to overestimate the importance that relational databases have attained in our lives. (For example, relational databases are at the heart of e-commerce.) We will look a bit further at relational databases towards the end of this chapter. □

4.1.4 Using the Cartesian Product

A relation can also be specified as a subset of the Cartesian product of the domain and codomain sets. To see why this works, let's briefly revisit the

Cartesian product, which we first saw in Chapter 1. Consider the sets

$$A = \{\text{sky, water, earth}\} \qquad \text{and} \qquad B = \{\text{blue, gray, brown}\}$$

The Cartesian product of these two sets, which is written as $A \times B$, is the set of all ordered pairs of the form (x, y) where x comes from A and y comes from B, i.e.,

$$\begin{aligned} A \times B = \{&(\text{sky, blue}), (\text{sky, gray}), (\text{sky, brown}),\\ &(\text{water, blue}), (\text{water, gray}), (\text{water, brown}),\\ &(\text{earth, blue}), (\text{earth, gray}), (\text{earth, brown})\} \end{aligned}$$

If you compare this to the examples that we have looked at so far, you can see that the Cartesian product itself can in fact be interpreted as a relation! The Cartesian product is the relation that contains *all possible connections* between the elements of its domain and codomain. Any other relation between these two sets will contain either the same number of ordered pairs or a smaller number of pairs. Consider the relations

$$\begin{aligned} \text{Winter} &= \{(\text{sky, gray}), (\text{water, gray}), (\text{earth, brown})\}\\ \text{Summer} &= \{(\text{sky, blue}), (\text{water, blue}), (\text{earth, brown})\} \end{aligned}$$

Each of these relations has been defined by listing out all of its connections between the domain A and codomain B as a set of ordered pairs. For example, consider the relation Winter, which has some of the elements of the Cartesian product $A \times B$ and no elements that are not in $A \times B$. Therefore we see that Winter is a *subset* of $A \times B$, which you'll recall we write as Winter $\subseteq A \times B$.

You can see by inspection that the same is true for Summer, so we can write Summer $\subseteq A \times B$. In fact any relation r between A and B has to be a subset $r \subseteq A \times B$. When a relation is specified as a list of ordered pairs (as we did for Winter and Summer) then there is no need to specify the rule for connecting elements (as we did in Section 4.1.1), because the rule is implicit in the list.

In fact there doesn't have to be any obvious rule for connecting the pairs at all! The relation

$$\{(\text{sky, brown}), (\text{water, brown}), (\text{sky, gray}), (\text{ground, blue})\}$$

is a perfectly valid relation, even though the ordered pairs in it seem to be senseless when considered with Summer and Winter above.

Conversely, we can also define a relation by specifying the rule in terms of the Cartesian product. This is especially useful for relations that would be too big to list out each and every element. We use set builder notation described in section 1.2 to specify the rule. Consider the following example.

Example 4.10: Write out the list of elements in the relation

$$r = \{ (x, y) \in \mathbb{N} \times \mathbb{N} : x = y + 1 \}.$$

Solution: The list of members of r can be written out as

$$r = \{(1, 0), (2, 1), (3, 2), \ldots\}.$$

Since the set is infinite, we write down the first few elements (enough that the pattern is clear) and then use the ellipsis. □

Of course, this relation r is the same as that in Example 4.3, but now the rule has been specified in set builder notation rather than in English. This has the advantage of being more precise.

Example 4.11: Write out the list of elements in the relation

$$r = \{ (x, y) \in \mathbb{N} \times \mathbb{N} : x + y \text{ is odd} \}.$$

Solution: To write out the list of elements, we need first to think a little about what you get when you add even and odd numbers. The sum of two odd numbers is always even, and the sum of two even numbers is also even.

In fact, an odd number is only produced when an even number and an odd number are added together, so we just have to list these out systematically for the first few cases and then resort to using an ellipsis! The list of members of r can be rewritten out as

$$r = \{(0, 1), (1, 0), (1, 2), (2, 1), (2, 3), (3, 2), \ldots\}.$$ □

Example 4.12: List the relation on $\mathbb{N}$ defined by

$$r = \{ (x, y) \in \mathbb{N} \times \mathbb{N} : x + y \text{ is even} \}.$$

as a subset of the Cartesian product $\mathbb{N} \times \mathbb{N}$.

Solution: The relation will contain all pairs of numbers in which the first number equals the second, all pairs where both are odd, and all pairs where both are even. (In fact the first case is covered by the second two cases—but we list it separately here just for purpose of exposition.)

$$\begin{aligned}
\{&(0,0),(1,1),(2,2),\ldots,\\
&(1,3),(1,5),(1,7),\ldots,\\
&(3,5),(3,7),(3,9),\ldots,\\
&\ldots,\\
&(0,2),(0,4),(0,6),\ldots,\\
&(2,4)(2,6),(2,8),\ldots,\\
&\ldots\}
\end{aligned}$$

□

4.2 Properties of Relations

When you have worked out a few of the examples in the previous section, you will notice that relations tend to fall into several classes based on the pattern of their connections. Some of these patterns are useful. For example, the relation "is in the same family as" on the set of all people has an interesting property called transitivity. We could make use of this property if we were building a social networking web site—it allows us to link together people who have family members in common. In this section, we introduce three useful patterns of connections, called *reflexivity*, *symmetry* and *transitivity*. In Chapter 5, we will see additional ways to classify relations. Since we have developed several ways to specify relations (in English, by picture, by table and mathematically) this section will indicate how to identify whether a relation conforms to the three patterns in each of those ways.

4.2.1 Reflexivity

- In English:
  A relation is *reflexive* if and only if it contains all ordered pairs that have the same element in both positions, e.g., (x, x) for all elements x.

- By picture:
  A relation is *reflexive* if and only if there is a *loop* at every node, i.e., an arrow starting at that node and going directly back to itself. Figure 4.3 shows an example of a relation that is reflexive (on the left) and one that is not reflexive (on the right).

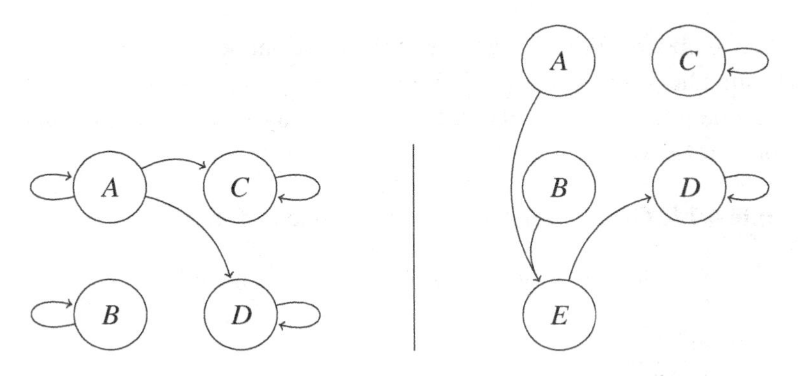

Figure 4.3: Examples of reflexive (left) and non-reflexive (right) relations.

- By table:
  A relation is *reflexive* if and only if for every element of the domain there is a column in which both rows have the same value.

- Mathematically:
  If r is a relation on A, then we say that r is *reflexive* if and only if

$$(x, x) \in r \qquad \text{for all } x \in A.$$

To save a bit of space, we can make use here of a notation that was introduced in section 3.2 on predicate logic. When we want to indicate that a proposition is true for all members of a set, we will use the *universal quantifier* symbol ∀, instead of writing "for all." We can rewrite the definition of reflexivity using this symbol as follows:

$$(x, x) \in r \qquad \forall x \in A.$$

Example 4.13: Determine whether the relation r on the set of people in your high school graduation class, specified as

(x, y) is in the relation r if person x is the same height as person y,

is reflexive or not.

Solution: Each reader's high school graduation class will probably have had a different distribution of heights. Nonetheless, we can still say with confidence whether the relation r is reflexive even without knowing who was in the class. The main question to consider is: if person x is in the class then is (x, x) in the relation? (This comes straight from the definition of reflexive above.) For (x, x) to be in the relation, person x would have to be the same height as person x. But this is always true—a person is always the same height as herself! Therefore we can say with confidence that the relation is reflexive. □

Example 4.14: Given a relation r on $\mathbb{N}$ defined as follows

$$r = \{ (x, y) \in \mathbb{N} \times \mathbb{N} : (x + y) \text{ is even} \}.$$

Is r reflexive?

Solution: Unlike the previous example, it would indeed be possible to work out the list of ordered pairs in r, but it would be a long list. Instead we think about what it takes to get an even (or odd) number:

- The sum of two even numbers is even.
- The sum of two odd numbers is even.
- The sum of an even number and an odd number is odd.

The main question to consider is: if we have a natural number x then is (x, x) in the relation? If x is even then, yes, two even numbers added will be even, so x is in the relation. If x is odd, then two odd numbers added will also be even, so again x is in the relation. But x can only have been odd or even, there are no other choices. So no matter what x we pick, (x, x) is in the relation. Hence, the relation is reflexive. □

Example 4.15: Is the relation r on the set $\mathbb{Z}$ of integers defined below reflexive?

$$r = \{ (x, y) \in \mathbb{Z} \times \mathbb{Z} : x > y \}$$

Solution: To show that something is not reflexive, it's sufficient to show that there is at least one x for which (x, x) is not in the relation. So picking any number in $\mathbb{Z}$, such as $x = 3$, we ask is $(3, 3)$ in the relation r? We can be sure that $(3, 3)$ is not in the relation because 3 is not bigger than itself (no number is). Hence this relation is not reflexive. □

Warning: Suppose that a relation has absolutely no (x, x) members. We then say that this relation is *irreflexive*. Be careful to distinguish this from non-reflexive, which just means that (x, x) is not present for *all* x (even though there may be some x for which it is present).

The relation r above is non-reflexive *and* it's also irreflexive. □

Example 4.16: [SN] Do you think that the "friend" relation in most social networking sites is reflexive?

Solution: No, it is not, because most social networking sites do not allow you to "friend" yourself. (Unless, of course, you set up multiple accounts.) But even if it were possible, the relation would still not be reflexive unless you were obliged to *always* "friend" yourself. It is unlikely that any social networking site would be set up this way. Certainly Facebook is not. □

4.2.2 Symmetry

- In English:
  A relation is *symmetric* if when it contains a given ordered pair (x, y), it also contains the mirror image (y, x), i.e., the same pair, but in the reversed order.

- By picture:
  A relation is *symmetric* if it has no "one-way streets." Figure 4.4 shows an example of a relation that is symmetric (on the left) and one that is not symmetric (on the right).

- By table:
  A relation is *symmetric* if and only if for every column with rows x and y, there is another column with rows y and x.

- Mathematically:
  If r is a relation on a set A, then we say r is *symmetric* if and only if

  $$(x, y) \in r \implies (y, x) \in r \qquad \forall x, y \in A.$$

  where $\implies$ is the logical implication operation introduced in the chapter on Logic.

Once again, we use "$\forall x, y \in A$" to mean "for all values of x and of y in the set A."

Example 4.17: Given a relation on the set of people in your high school graduation class defined by

(x, y) is in the relation if person x is the same age as person y.

Is this relation symmetric?

Solution: We cannot tell who was in your high school graduation class, of course. Nonetheless, we can determine some general facts about their

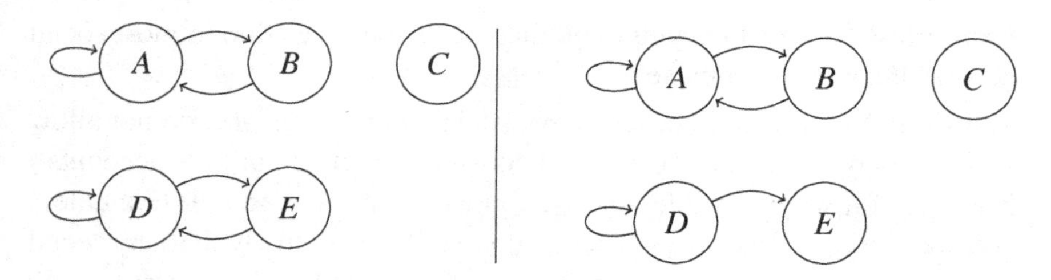

Figure 4.4: Examples of symmetric (left) and non-symmetric (right) relations.

ages. The main question to ask (from the definition of symmetry) is: if you are the same age as someone else, is she the same age as you? Of course she is! This is sufficient for us to conclude that the relation above is symmetric. □

Example 4.18: Given a relation r on $\mathbb{N}$ defined by

$$(x, y) \in r \iff x + y \text{ is even.}$$

Is r symmetric?

Solution: Note that the rule for the relation in this example is written a little differently; it says that (x, y) is in the relation r if and only if $x + y$ is an even number. We could have written this as a subset of the Cartesian product as

$$r = \{\, (x, y) \in \mathbb{N} \times \mathbb{N} : x + y \text{ is even} \,\},$$

and this would be the same relation (i.e., the same set of ordered pairs).

It would not be possible to write out all the elements of this relation and then check the definition. However, some reasoning will suffice instead. When we add two numbers x and y and we get an even result, then if we were to add the numbers in reverse order (y and then x) would we still get an even result? Of course we would—the order does not matter in addition (i.e., addition is *commutative*). Therefore this relation is symmetric. □

Example 4.19: Given a relation r on $\mathbb{Z}$ defined by

$$(x, y) \in r \iff x > y.$$

Is r symmetric?

Solution: To show that something is not symmetric, it's sufficient to show that there is at least one x and y for which (x, y) is in the relation but

(y, x) is not. Let's pick a pair in this relation, say $(3, -2)$; is $(-2, 3)$ in the relation? Well, since -2 is not greater than 3, we can be confident that $(-2, 3)$ is not in the relation. As all we had to find was one missing case, we can confidently say that the relation is not symmetric. □

In Section 4.2.1, we looked at both reflexive relations and their diametrical opposites (irreflexive relations). It will be useful to do the same with symmetric relations. Suppose a relation has the following property: if (x, y) belongs to the relation, where $x \neq y$, then (y, x) *never* belongs to the relation. We say that such a relation is *antisymmetric*. Graphically, this means that *all* streets are one-way streets; however, loops are allowed (remember the condition $x \neq y$). Be careful to distinguish this from a non-symmetric relation, which just means that (y, x) is not present for all (x, y) (even though there may be some (x, y) for which it is present). In the graph of a non-symmetric relation, we may wind up finding both one-way streets and two-way streets.

Warning: Don't fall into the trap of thinking that all relations must be either symmetric or antisymmetric! Many relations are neither symmetric nor antisymmetric. We shall see one such relation when we study Example 4.23 below. □

Example 4.20: Is the relation $\geq$ on $\mathbb{Z}$ symmetric? antisymmetric? neither?

Solution: For symmetry, we must ask ourselves the question, "If $x \geq y$, is $y \geq x$?" Let us pick an example, say 10 and -2, both integers. Clearly $10 \geq -2$ but just as clearly -2 is not greater than or equal to 10. Since we have found one case that does not obey the rule for symmetry, the relation is not symmetric.

Is $\geq$ antisymmetric? Suppose that x and y are distinct (i.e., different) integers such that (x, y) belongs to the $\geq$ relation. If it is always the case that (y, x) does not belong to the $\geq$ relation, then the $\geq$ relation is antisymmetric.

So let's check. We are told that (x, y) belongs to the $\geq$ relation, so $x \geq y$; we are also told that x and y are distinct, so $x \neq y$. The statement that $x \geq y$ means that either $x > y$ or $x = y$, but the latter has already been ruled out, since x and y are distinct. Thus $x > y$, which means that $y > x$ is false. We are almost there. Since $y \not> x$ and $y \neq x$, we see that $y \geq x$ is false.

In short, we have shown that if x and y are any distinct integers such that (x, y) belongs to the $\geq$ relation, then (y, x) does not belong to the $\geq$ relation. Thus $\geq$ is antisymmetric. □

Example 4.21: Determine if the relation "x is married to y," on the set of all people in the U.S.,[1] is symmetric, antisymmetric or neither.

Solution: If person a is married to person b, can it ever be the case that person b is *not married* to person a? Since legal marriage is an equal partnership between two people, it must be a symmetric relation. □

To save time, we sometimes simply list the pairs that form a relation on a set, without explicitly naming the set. In that case, we determine the domain by inspection as simply being all the elements that appear in the relation. The next example should clarify what this means.

Example 4.22: Determine if the relation

$$r = \{(3, 4), (4, 4), (2, 3), (4, 3), (1, 2), (2, 1), (3, 2)\}$$

is symmetric, antisymmetric or neither.

Solution: This relation is described as a list of ordered pairs. By inspection, the smallest set that the relation can be on is the set $\{1, 2, 3, 4\}$. (We arrived at this conclusion by collecting the first element of each ordered pair into a set.) When we don't explicitly specify the set that a listed relation is on, we will consider it fair to assume that it is the smallest set that can be created from the first element in each ordered pair. In that case, also by inspection, we can tell that whenever (x, y) is a pair in the relation, then (y, x) is there as well. So the relation is symmetric. □

The smallest set assumption is important. r could also be a relation on the set $\{1, 2, 3, 4, 99\}$. But in that case, r would *not* be symmetric.

Example 4.23: Determine if the relation "knows the birthday of," on the set of all people, is symmetric, antisymmetric or neither.

Solution: I know the birthday of the Queen of the Netherlands (it's a national holiday in the Netherlands); however, she is quite unaware of my birthday (she never sends me a card). The relation is not symmetric.

On the other hand, I know my wife's birthday and she knows mine. The relation is not therefore antisymmetric.

So, this relation is neither symmetric nor antisymmetric. □

Example 4.24: [SN] In Facebook, is the "friend" relation symmetric, antisymmetric or neither?

Solution: The relation is symmetric. When you "friend" a person in Facebook, they are sent a friend request from you. When they accept this friend

[1] The rules of marriage can differ from country to country.

request, then you are each recorded as a friend of the other. So in Facebook, the "friend" relation is constructed so as to be symmetric. □

4.2.3 Transitivity

The reflexivity property of a relation on A can be determined by just looking at all single elements of A. The symmetric property of a relation on A requires that you look at all pairs of elements of A. The transitive property of a relation on A will require inspection of all triples of elements of A.

- In English:
  A relation is *transitive* if it contains ordered pairs that allow "shortcuts." That is, for a relation to be transitive, we mean that if (x, y) and (y, z) are both in the relation, then it must be possible to also take a shortcut directly from x to z, that is, (x, z) must also be in the relation.

- By picture:
  Figure 4.5 shows an example of a relation that is transitive (on the left) and one that is not transitive (on the right). The example on the left has the property that whenever we can travel from one circle to a second, and then on to a third, there will *always* be a shortcut from the first directly to the third!

- By table:
  A relation is *transitive* if and only if whenever there is a column with rows x and y, and a column with rows y and z, then there is also a column with rows x and z.

- Mathematically:
  If r is a relation on the set A, then we say $r \subseteq A \times A$ is *transitive* if and only if

$$(x, y) \in r \wedge (y, z) \in r \implies (x, z) \in r \qquad \forall x, y, z \in A.$$

Example 4.25: Given a relation on the set of people in your high school graduation class defined by the rule

(x, y) is in the relation if x the same height as y.

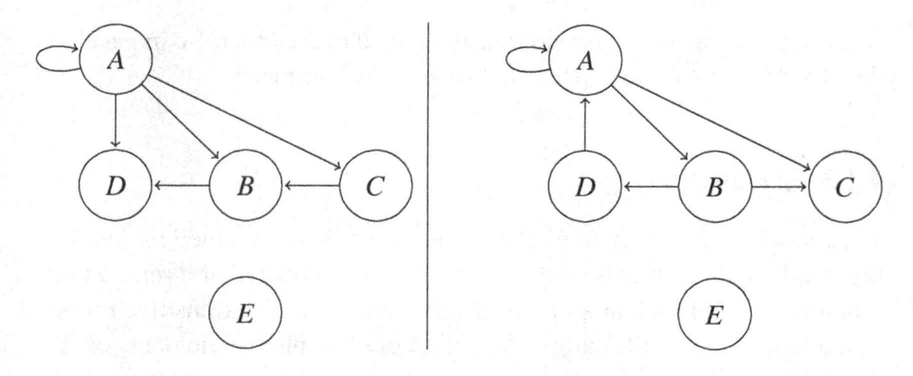

Figure 4.5: Examples of transitive (left) and non-transitive (right) relations
.

Is this relation transitive?

Solution: Let's consider some people in the class: if (x, y) is in the relation, then person x is the same height as person y. If (y, z) is in the relation then person y is the same height as person z. In all graduating classes where we see this pattern, are we guaranteed that (x, z) is in the relation—that is, are we guaranteed that person x is the same height as person z? When phrased in English like this, it should be clear to you that this always has to be the case.

Note that for the purpose of testing transitivity, the "the same height as" relation is essentially the same as testing the "is equal to" operation on the set of real numbers. □

Warning: In some high school graduating classes, we may find that (x, y) is in the relation but there is never a (y, z). That is, we never have more than two people of the same height. However, the definition of transitive can only be tested for if there is a (y, z). Strangely enough, if there are no (y, z) cases ever in the relation we are testing, then the relation is also considered transitive. For this reason, it is best to consider a relation to be transitive until you can show that it is missing a "shortcut." □

Example 4.26: Given a relation on $\mathbb{N}$ defined by the rule

$$r = \{ (x, y) \in \mathbb{N} \times \mathbb{N} : x + y \text{ is even} \}.$$

Is this relation transitive?

Solution: The question we should ask is: if $x + y$ is even and $y + z$ is even, then does it have to be the case that $x + z$ is even? You'll recall our reasoning that x and y must be both even or both odd for their sum $x + y$ to be even. Let's consider both cases:

- Suppose that both x and y are even. Since y is even and $y + z$ is even, we see that z must be even as well. In that case $x + z$ is the sum of two even numbers, which is even.

- Of course the same reasoning applies to both x and y being odd, in which case z would have to be odd and again $x + z$ will be even.

Therefore, we have shown that if both (x, y) and (y, z) are in the relation, then so is (x, z). Hence the relation is transitive. □

Example 4.27: Given a relation on $\mathbb{Z}$ defined by the rule

$$r = \{ (x, y) \in \mathbb{Z} \times \mathbb{Z} : x \text{ is 10 greater than } y \}.$$

Is this relation transitive?

Solution: By now you may have noticed that we always have the same approach to showing that a relation does not have a property: we find a case of the property in the relation that doesn't work! So, let's pick some numbers: 50 is 10 greater than 40, so $(50, 40)$ is in the relation. Since 40 is 10 greater than 30, we have that $(40, 30)$ is in the relation. Now we can ask: is $(50, 30)$ in the relation? Since 50 is 20 greater than 30 it is not in the relation. Hence the relation is not transitive. □

Example 4.28: Is the relation $\leq$ on $\mathbb{Z}$ transitive?

Solution: Let's pick some pairs in this relation: $-2 \leq 4$ so $(-2, 4)$ is in the relation. Also $4 \leq 8$ so $(4, 8)$ is in the relation. Is $(-2, 8)$ in the relation? Since $-2 \leq 8$, then yes, it is in the relation. We can repeat this for some other pairs to give ourselves some measure of confidence that this is a transitive relation. Then we can ask ourselves for (x, y) and (y, z) in the relation: what would it take for (x, z) not to be in the relation? Since $x \leq y$ and $y \leq z$ for (x, z) not to be in the relation we would need x not to be less than or equal to z; that is for x to be greater than z. But that can never happen if y is a value between x and z and y is smaller than or equal to z. The relation is transitive. □

Example 4.29: Consider the relation "is taking the same class as" defined on the set of people in this University. Is this relation transitive or non-transitive?

Solution: Let's pick some example pairs: Pete might be taking the same class as Joey. Joey might be taking the same class as Carrie. However, since these might be different classes, Carrie might, or might not, be taking the same class as Pete. Hence the relation is not transitive. □

Example 4.30: **[SN]** In Facebook, is the "friend" relation transitive?

Solution: In Facebook, the "friend" relation is not transitive. Let's consider that Alexis is one of your friends on Facebook. For the "friend" relation to be transitive, every one of Alexis' friends would have to also be your Facebook friend as well. And this would have to be true for all your friends, not just for Alexis. While this is, of course, possible, it's not required or even likely. Hence the "friend" relation is not transitive. □

4.3 Relational Databases

The idea of storing information on a computer in a relational database was first proposed by Edgar F. Codd of IBM in the 1970s. Modern relational databases (sometimes written RDMS, an acronym that stands for Relational Data Base Management Systems) include *Oracle*, *Microsoft Access* (small applications), *Microsoft* SQL *Server* (bigger applications) , and IBM's *dBase*. These tools are at the core of the information revolution and it's important for professionals in all fields to understand them beyond a rudimentary level.

Information is stored in a relational database in tables similar to those in the Appendix. Tables can have any (finite) number of columns. However, one column is special and is called the *key*. This column is the unique index into the table. In Table A.1, the key is the *Name* column. Any database table needs to have a key column.[2] In our example here, we have simplified the situation so that the name is unique and hence can be used as a key. For a real social networking site database, an email address, might be a better key than a name, because it's more unique. The other columns provide attribute information about the entry in the key column. One of the advantages of this tabular approach is that not all tables have to have information for every key value. For example, Lena appears in Table A.1. However, we don't have any educational information for Lena—she chose not to fill that in for her profile. Nonetheless we do not have to waste space with "empty" entries for her educational information; although she

[2]This is a simplification.

appears in the basic table, she simply does not appear in the educational information in Table A.2

Example 4.31: [SN] Determine which columns of Table A.2 and Table A.3 will qualify as as key columns.

Solution: No columns in these tables can be used as key columns because every column has at least one repeated value. It is, however, straightforward to modify the tables to have a unique key by simply adding a column that contains the row number. □

Information is extracted from a relational database by looking up tables for key entries, or by looking across several tables and cross-indexing the information. This combination, simple as it seems, is very powerful. For example, we can look up the entry for the key value *Name* = *Sandra* in Table A.1 and extract the birthdate 08/03/1967. We can also cross-index the *Name* in the Table A.2 and extract 1988 as the year of graduation. Together these tell us that Sandra was 21 when she graduated from Columbia, a fact not explicitly stored anywhere.

The most common way to extract information from a relational database system is to use a special language called the *Structured Query Language* or SQL, a language invented very soon after the relational database in the 1970s. SQL[3] has since come to be adopted as an international standard and it would be rare for a relational database system not to support queries (that is, asking questions to extract information) in SQL. The formal framework that SQL is based upon is *relational algebra*, a combination of relations and logic.

The **select** operator in SQL will extract information from a table. Let's denote Table A.1 by the name *Friends*. The following SQL query will extract the names and sex of all the married people in the table:

select *Birthdate, Sex*
from *Friends*
where *Status* = M;

The background we have build up in sets, logic and relations allows us to now see how this has to work at a more fundamental level.

Example 4.32: Show how the SQL **select** example above can be represented as a relation.

---

[3]Which is most often pronounced "sequel," but sometimes "es-queue-ell."

Solution: Let us suppose that we have represented Friends as a 6-ary relation as we did in Example 4.8. In that case, we have

$$r_\mathrm{f} = \{(n_i, c_i, h_i, s_i, b_i, st_i) : i \in 0, \ldots, 25\}.$$

where i is just the row number in Table A.1 and where the entries in the 6-tuple are just the column values in order. The result of the SQL **select** example above can be written very simply as the set

$$\{(b_i, s_i) : (n_i, c_i, h_i, s_i, b_i, st_i) \in r_\mathrm{f} \wedge st_i = \mathrm{M}\}. \qquad \square$$

Let's see now how you can cross-reference tables in SQL. In this example, we want to extract for everybody the information that we extracted for Sandra, that is, her birthdate and graduation date. Let's refer to Table A.2 as *Education*.

select *Name, Birthdate, Class*
from *Friends* **join** *Education*
on *Friends.Name* = *Education.Name*;

The result of this **join** operation is a table with three columns (*Name*, *Birthdate* and *Class*) made by cross-referencing between the *Friends* and the *Education* tables using the *Name* column as the common lookup field.

Example 4.33: Show how the SQL **join** operation given above can be represented as a relation.

Solution: We already developed a relation to represent the table *Friends* in Example 4.32. We represent the table *Education* as a 5-ary relation as follows:

$$r_\mathrm{e} = \{(n_i, u_i, c_i, d_i, m_i) : i \in 0, \ldots, 19\}.$$

The new relation that is the result of the SQL query will contain some elements from each of r_f and r_e, namely those elements with a common value in their name fields

$$\{(n, b_i, c_j) : (n, c_i, h_i, s_i, b_i, st_i) \in r_\mathrm{f} \wedge (n, u_j, c_j, d_j, m_j) \in r_\mathrm{e}\}. \qquad \square$$

In our final example, let's see how the properties of a relation might be used by a social networking company. Consider the friendship relation as shown in Figure A.1. The "friend" relation on a social networking set is not transitive, as we have discussed earlier in this chapter. However, if you are a member of a social networking site, then you might indeed be interested to know the friends of your friends—some of them may be folks whom you have met and with whom you are friendly. Perhaps you would have friended these folks on your social networking site if only you knew they had an account! In Figure A.1, Grace is a friend of Rania. A clever social networking site might propose some of Rania's friends as possible new friends for Grace (without saying they were Rania's friends and hence disclosing the source). Let's see if we can write some SQL code to do this.

Example 4.34: Given a two column table *FriendOf* (with columns *Name1* and *Name2*) that captures the relation in Figure A.1, write SQL code that uses transitivity to find new friend suggestions.

Solution: We will need two copies of the table. Let's call one *FriendOfA* and the other *FriendOfB*. The following SQL query identifies all the transitive "shortcuts" that are the "friends of friends" for every person in the network.

```
select FriendOfA.Name1, FriendOfB.Name2
  from FriendOfA join FriendOfB
  on FriendOfA.Name2 = FriendOfB.Name1
  as FriendSuggestions;
```

The two copies of the table are cross-indexed by the second to last line, so that for every person in table A (that is for every *FriendOfA.Name1*) that person's friends are looked up in table B (*FriendOfA.Name2 = FriendOfB.Name1*) and all *their* friends (*FriendOfB.Name2*) are then placed into a new table called *FriendSuggestions* under the person's name. □

4.4 Exercises

4.4.1 Explain why (Sandra, Molly) is also in the relation of Example 4.2 and why Mark won't ever feature as the first element in a relationship pair.

4.4.2 List the relation in Figure 4.1 as a subset of a Cartesian product. (Remember to use set notation consistently.)

4.4.3 Write out the set of ordered pairs in the following relations on the natural numbers $\mathbb{N}$:

(a) (x, y) is in the relation if $x < y$

(b) (x, y) is in the relation if $x + 1 = y + 2$

(c) (x, y) is in the relation if $2x + y$ is odd.

(d) (x, y) is in the relation if $x - y$ is odd.

(e) (x, y) is in the relation if $x^2 + y$ is odd.

4.4.4 List the relation on $\mathbb{N}$ defined by

$$r_b = \{ (x, y) \in \mathbb{N} \times \mathbb{N} : x > y \}$$

as a subset of the Cartesian product.

4.4.5 Write out the following relations as subsets of the Cartesian product:

(a) The relation on $\mathbb{N}$ defined by

$$r_c = \{ (x, y) \in \mathbb{N} \times \mathbb{N} : x = y \}.$$

(b) The relation on $\mathbb{N}$ defined by

$$r_d = \{ (x, y) \in \mathbb{N} \times \mathbb{N} : x = y + 5 \}.$$

4.4.6 Determine if the following relations are reflexive, irreflexive or neither.

(a) The relation $x < y$ on the set $\mathbb{Z}$.

(b) The relation "x is married to y" with domain and codomain the set of all people in the USA.

(c) The relation $r_1 = \{ (x, y) \in \mathbb{N} \times \mathbb{N} : x - y = 1 \}$.

(d) The relation $r_2 = \{(3, 4), (4, 4), (2, 3), (3, 3), (1, 2), (2, 1), (4, 3)\}$.

4.4.7 Determine if the relation $r_1 = \{ (x, y) \in \mathbb{Z} \times \mathbb{Z} : x + y = 1 \}$ is symmetric, antisymmetric or neither.

4.4.8 Consider the following relations on the set $\mathbb{N}$ of natural numbers and say if the resulting relation is: reflexive, irreflexive or neither; symmetric, antisymmetric or neither; and transitive or not.

(a) $\{(3, 4), (6, 8), (3, 9), (4, 3), (9, -2), (4, 9)\}$.

(b) $\{(2, 2), (1, -3), (-3, -3), (1, 1), (-3, 1)\}$.

(c) $\{(1, 1), (1, 2), (1, 3), (2, 3), (2, 4), (3, 4), (5, 5)\}$.

4.4.9 Consider the following relations and say whether they are transitive or not:

(a) The relation "x is married to y," with domain and codomain the set of all people in the USA.

(b) The relation $r_1 = \{ (x, y) \in \mathbb{N} \times \mathbb{N} : x - y = 1 \}$.

4.4.10 Consider the following rules on the set $\mathbb{N}$ of natural numbers and say whether resulting relation is: reflexive, irreflexive or neither; symmetric, antisymmetric or neither; and transitive or not.

(a) Is less than or equal to.

(b) Is a factor of.

(c) Is 3 less than.

4.4.11 Consider the following rules on the set of all people and examine the resulting relations and say if they are: reflexive, irreflexive or neither; symmetric, antisymmetric or neither; and transitive or not.

(a) Is smarter than.

(b) Went to the same high school as (don't forget about transfer students!).

(c) Is a child of.

(d) Is a cousin of.

(e) Visited a museum in common with.

(f) Has the same parents as.

4.4.12 For each of the following, determine whether they are: reflexive, irreflexive or neither; symmetric, antisymmetric or neither; and transitive or not.

(a) $r_1 = \{ (x, y) \in \mathbb{N} \times \mathbb{N} : x - y \text{ is odd} \}$.

(b) $r_2 = \{ (x, y) \in (\mathbb{R} - \{0\}) \times (\mathbb{R} - \{0\}) : x/y = 1.0 \}$.

(c) $r_3 = \{ (x, y) \in \mathbb{N} \times \mathbb{N} : 2x \geq y \}$.

(d) $r_4 = \{ (p, q) \in \mathbb{Z} \times \mathbb{Z} : p^2 + q \text{ is odd} \}$.

(e) $r_5 = \{ (a, b) \in \mathbb{N} \times \mathbb{N} : a + 2b \text{ is odd} \}$.

4.4.13 [SN] Write SQL code to extract information from the *Friend* and *Education* tables (that is, Table A.1 and Table A.2) as follows:

(a) The person's name and hometown for each of the males in the social network.

(b) The names of all the people who live in South Park.

(c) The name and hometown of everybody who studied Computer Science .

4.4.14 [SN] Using the *Friends* and *Education* tables (that is, Table A.1 and Table A.2), write down the relations that correspond to the following SQL operations

(a) **select** Name **from** *Friends* **where** *Sex* = M **and** *Status* = M;

(b) **select** *Degree* **from** *Education* **where** *Class* > 1990;

(c) **select** *Name* **from** *Friends* **join** *Education*
on *Friend.Name* = *Education.Name*
where *Education.Degree* = BA;

4.4.15 [SN] You work for a famous social networking company. The irascible CEO tells you that he wants you to expand the coverage of the company's social network by exploiting the relation "attended the same University."

(a) This relation may not produce quite as much growth as the CEO hopes—why not?

(b) You see an opportunity to shine, can you propose a modified relation that would work better?

(c) Assume tables such as *Friends* and *Education* in the previous two exercises and write SQL code that identifies potential new friends using this relation.

Chapter 5

Functions

> I believe that numbers and functions of Analysis are not the arbitrary result of our minds; I think that they exist outside of us, with the same character of necessity as the things of objective reality, and we meet them or discover them, and study them, as do the physicists, the chemists and the zoologists.
>
> Charles Hermite

We now turn to the topic of functions. On the one hand, functions may be thought of as "special" relations; on the other hand, functions may be thought of as the fundamental basis for all computation! Let's briefly address these two points, in turn.

In Chapter 4, we saw how relations can be used to represent structured knowledge. Such knowledge often takes the form of the relationships we encounter in our everyday life. For example, we can define a relation to represents the notion of "went to high school with" over a group of people. This relationship would connect any given person to each person with whom she attended high school, and thus we would not usually expect a person to be linked up with exactly one person in this relation. Indeed, if you attended a relatively large high school, you would be related to *many* people via this relationship. At the other extreme, somebody might be related to *nobody* in our set of people, say, if he were home-schooled and had no siblings.

So, we can think of a relation as providing a *set* of things that are related to a given item. Sometimes this set is empty; sometimes, it has two (or more) elements. But sometimes this set contains exactly one element. For example, suppose we were interested in the (last) high school attended by a person, where each of the people involved had attended (at least one) high school. The question "what was the last high school you attended?" only has one answer. Such a relationship is said to define a *function.*

So the main difference between general relations and functions is that relations are *multiple-valued* (e.g., we each went to high school with a large number of people), but functions are *single-valued* (e.g., we each graduated from exactly one high school).

As another example, consider your set of friends. It's a good bet that they all have email addresses. In fact, many of them will have multiple email addresses—one for work, one for home, one for their eBay business, one for their volunteer activities and so forth. But which is their favorite? That is, which one should you use to make sure they read your message? It would be useful if when you send a message to a friend on Facebook, it uses the friend's favorite address, rather than (say) a somewhat randomly-chosen address or the first address for your friend that you happened to enter. We can represent this need by using a *function* that *maps* each of the friends in your set of friends, to an email address in the set of emails of your friends; the *value* of this function (for a particular friend) is her favorite email address.

Both functions and relations appear in the implementation of social networking websites. For example, Facebook allows you to specify your gender (male or female), relationship status (single, in a relationship, engaged, married, "it's complicated," open relationship, widowed), and birthday (a specific choice of month, day, and year). All of these are functional relationships; you only have one gender, one relationship status, and one birthday. Note that you specify your choice via a drop-down menu, which forces you to make exactly one selection (radio boxes would've worked, as well). However, Facebook also allows you to specify non-functional relationships,[1] i.e., relationships that are not functions. Examples of the latter include who you're interested in (women, men) and what you're looking for (friendship, dating, a relationship, networking). Since these are data selected via checkboxes, you can choose any subset (perhaps empty) of same. For instance, if you chose "dating" and "networking", you would be connected to two different items within the "looking for" relation.

By this point, we have now explained (in general terms) how functions

[1]These should not be confused with dysfunctional relationships. ☺

are certain kinds of relations. What of our other claim, that functions underlie all of computation? As we shall see in Section 5.1, a function may be thought of as a mechanism (we'll call it a "black box") that turns inputs into outputs, with each possible input value being associated with *exactly* one output value. We can envision any computational task in this manner. Some examples:

- The Internet's Domain Name System (DNS): The input is the name[2] of a device on the Internet; the output is the numerical IP (Internet Protocol) address that uniquely identifies this device. The IP address is traditionally written in "dotted quad" form $a.b.c.d$, where $a, b, c, d \in \{0, 1, \ldots, 255\}$. (For example, the IP address corresponding to `www.dsm.fordham.edu` is `150.108.64.52`.)
- The World-Wide Web: The input is a Uniform Resource Locator (URL) describing a resource on the Web; the output is that resource itself.
- E-commerce: The input is an order for merchandise, to be submitted to an e-commerce website; the output is the sequence of actions that will cause the order to be delivered to the desired address.
- Scheduling: The input is a list of tasks to be done; the output is the best order in which they should be performed. Here, "best" generally means minimal-cost in some manner (e.g., time or money).
- Documents (of whatever kind, e.g., word processing, spreadsheet, artwork, music, etc.): The input is some kind of easily-editable representation of the document; the output is a version of the document that is suitable for some kind of output device. For example, this book was created using the LaTeX typesetting system. Input was the set of LaTeX commands that represents the book's structure and content; the final typeset output was represented as a PDF file.

The list of examples is only limited by your imagination.

5.1 What is a Function?

It is pretty likely that you've run across functions in earlier math classes that you have taken. You were probably asked to plot the graphs of curves,

[2]Technically speaking, the Fully Qualified Domain Name.

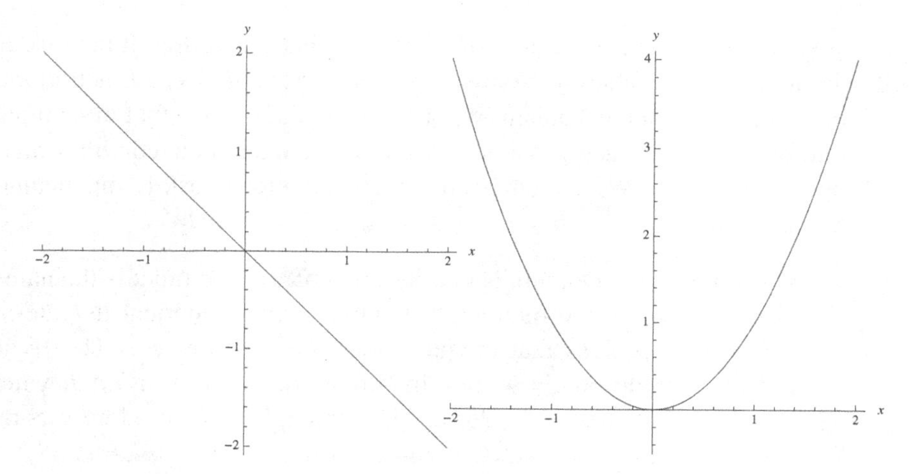

Figure 5.1: Graphs of the curves $y = -x$ and $y = x^2$

such as those found in Figure 5.1. Perhaps you have even been told that you were plotting "functions," rather than curves, but you probably weren't told exactly what this means.

We can explain a function as being a "black box,"[3] as in Figure 5.2. The main "parts of speech" that we see in Figure 5.2 are as follows:

- The set X denotes the set of all possible input values for the function. We call X the *domain* of the function.
- The set Y denotes the set of all possible output values for the function. We call Y the *codomain* of the function.
- The function is named f. You can think of this as representing the rule that tells us how to assign an output value to a given input value.

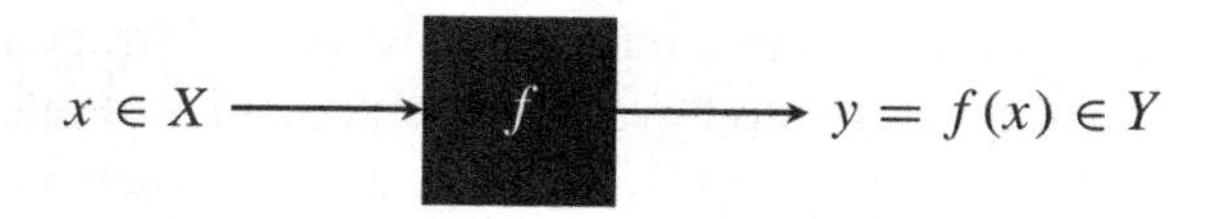

Figure 5.2: A function as a black box

We will often use the notation $f : X \to Y$ to mean that f is a function with domain X and codomain Y.

[3]The reason we draw this as a black box is that we often don't know (or care) how the function is computing its values. In other words, we can't (or don't) look inside the box to see how the function is operating.

So how do we go about describing a particular function? We have already seen that numerical functions can be described as graphs. Since numerical functions are only part of the story, we need other ways to describe functions. Let's recall what we know about describing relations; perhaps some of these ideas will also work for describing functions.

We can use English to describe functions. For example, the function on the left-hand side of Figure 5.1 can be described as the function

- whose domain is the set $\mathbb{R}$ of real numbers,
- whose codomain is $\mathbb{R}$, and
- which returns the output value $y = -x$ for a given input value x.

The function on the right-hand side of Figure 5.1 can be described as the function

- whose domain is the set $\mathbb{R}$ of real numbers,
- whose codomain is $\mathbb{R}$, and
- which returns the output value x^2 for a given input value x.

We can also describe functions via *tables*, which may be thought of as a compact way of listing all the function's values. This is somewhat more convenient than an English description, since tables are more concise than sentences in English.

Example 5.1: Here is a function

$$d\colon \{1, 2, 3, 4, 5\} \to \mathbb{N},$$

whose table is given by

t	1	2	3	4	5
$d(t)$	2	4	6	8	10

.

If you understand our notation, you'll see that the domain of d is the set $\{1, 2, 3, 4, 5\}$ and the codomain is the set $\mathbb{N}$ of natural numbers. If you're wondering why we chose to name this function "d", it's simply because "d" is the first letter in the word "double".

Note that the table compactly represents the listing

$$\begin{aligned} d(1) &= 2 \\ d(2) &= 4 \\ d(3) &= 6 \\ d(4) &= 8 \\ d(5) &= 10 \end{aligned}$$

of the values of d.

We'll now present a different function that has the same table as d. This will be the function

$$d^*: \{1, 2, 3, 4, 5\} \to \{2, 4, 6, 8, 10\}$$

whose table is

z	1	2	3	4	5
$d^*(z)$	2	4	6	8	10

As you can see, the domain of d^* is the set $\{1, 2, 3, 4, 5\}$ and the codomain of d^* is the set $\{2, 4, 6, 8, 10\}$. Note that d and d^* have the same domain and the same table, but they have different codomains. This explains why we have given them different names. □

Remark. In the spirit of Example 5.1, we could've also looked at the function $d^{**}: \mathbb{Z}^+ \to \mathbb{Z}^+$ given by the table

z	1	2	3	4	5	6	7	8	9	10	...
$d^*(z)$	2	4	6	8	10	12	14	16	18	20	...

Of course, we now have an infinite table, which you might find a bit disconcerting. But this should not bother you any more than the sequences we discussed in Section 2.1. In fact, looking at Example 2.1 on page 46, we see that we that we can describe d^{**} a bit more economically, by saying that $d^{**}: \mathbb{Z}^+ \to \mathbb{Z}^+$ given by the rule

$$d^{**}(x) = 2x \qquad \forall x \in \mathbb{Z}^+.$$

Here, we use a "shorthand notation" that was introduced in Section 3.2.1, writing the symbol $\forall$ as a shorthand notation for the phrase "for all." □

When the domain of a function is small, as in Example 5.1, a table might be the most convenient way to describe a function. On the other hand, there are many situations in which we really don't have an alternative to using a table. This typically occurs when there is no obvious pattern that gives us a rule for computing function values.

Let's look at some more examples of functions that might arise from situations that might occur during the course of a day:

Example 5.2: A coffee shop sells small, medium and large cups of coffee, with the following prices

c	$p(c)$
small	\$1.25
medium	\$2.15
large	\$2.75

This table describes a function p (for "price"). The domain is the set of all coffee sizes {small, medium, large} and the codomain is the set of all possible prices. The rule for this function is that $p(c)$ denotes the price of a given coffee size c. □

As you can see, the table giving the rule that defines a function can in either a horizontal layout (as in Example 5.1) or vertical layout (as in Example 5.2). You should choose whichever layout seems more convenient or natural to you.

Example 5.3: A bakery sells various breakfast items, with the following prices:

i	$b(i)$
bagel	\$1.00
croissant	\$1.25
danish	\$2.25
muffin	\$1.50

This table describes a function b (for "bakery"). The domain is the set of all breakfast items {bagel, croissant, danish, muffin} and the codomain is the set of all possible prices. A vertical layout seemed more natural to me than a horizontal layout; maybe that's because it reminds me of a menu. □

Example 5.4: My address book contains email addresses for all my friends. We can represent my address book as a table. Here might be a portion of this table:

n	$e(n)$
⋮	⋮
Harry Q. Bovik	`bovik@cs.cmu.edu`
James T. Kirk	`kirk@starfleet.federation.gov`
Darth Vader	`vader@empire.gov`
⋮	⋮

This table describes a function e (for "email"). The domain is the set of names for all my friends, and the codomain is the set of all possible email addresses. Note that it would be *very* inconvenient to write this table using a horizontal layout. □

Example 5.5: As mentioned earlier, each Facebook member specifies his or her gender. This can be thought of as a function $g\colon F \to G$, where F is the set of Facebook members, and $G = \{\text{M}, \text{F}, \text{U}\}$ is the set of gender choices (male, female, unspecified).[4] Assuming that the authors of this book are all Facebook members, we could use the table

p	$g(p)$
⋮	⋮
Bovik, Harry Q.	U
Lyons, Damian M.	M
Weiss, Gary M.	M
Papadakis-Kanaris, Christine	F
Werschulz, Arthur G.	M
⋮	⋮

to (partially) specify the function g. Note that Harry hasn't yet specified his gender choice; he's too busy or just can't be bothered. □

5.2 Functions and Relations

Note that a function can be described by a domain, a codomain, and a rule that tells you how to evaluate the function. Recall that a relation can be described by a domain, a codomain, and a rule that tells you which pairs of elements belong to the relation. So it appears that functions are similar to relations. What distinguishes between them?

[4]Facebook members need not specify their gender. In fact, they don't need to specify *any* personal information.

Go back to the graphs of the two functions that appear in Figure 5.1 on page 152 at the beginning of this chapter. Now a graph is a set of points (x, y) in the plane. If you look at the graphs of these functions, you'll see that for each possible x, there is exactly one y such that point (x, y) appears on said graph; that value of y is what we mean when we talk about *the* value of the function at the particular value x.

On the other hand, consider Figure 5.3, which is the graph of the curve given by $y^2 = x$, where x can be any non-negative real number. This is

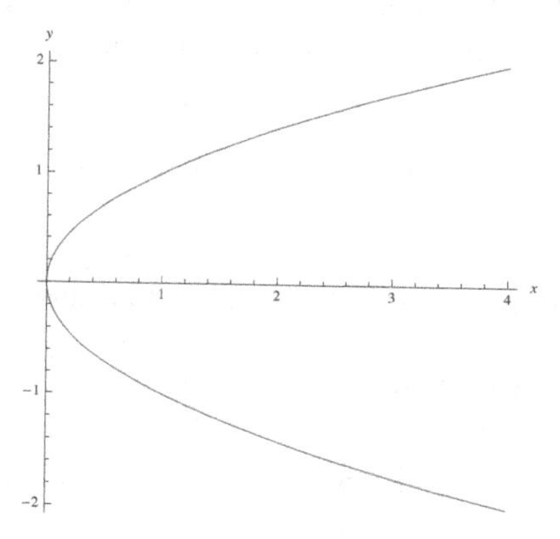

Figure 5.3: The curve $y^2 = x$

the graph of a relation, whose domain is the set $\mathbb{R}^{\geq 0}$ of non-negative real numbers and whose codomain is the set $\mathbb{R}$ of real numbers. Figure 5.3 is *not* the graph of a function. To see this, let $x = 4$. There are two y-values such that the point $(4, y)$ is on the curve $y^2 = x$, namely $y = 2$ (since $4 = 2^2$) and $y = -2$ (since $4 = (-2)^2$). As there is not a unique choice of y such that $(4, y)$ is on the curve, the curve does *not* define a function.

So it's clear that when distinguishing functions from relations, we need to disallow graphs such as Figure 5.3. Now both the graph of a function and the graph of a relation will always consist of a set of ordered pairs (x, y). The main difference between the graph of a function and a graph of a relation is as follows:

- If $f : X \to Y$ is a function, then for any $x \in X$, we must be able to talk about the $y = f(x) \in Y$ that corresponds to x.

- For a relation that is not a function, we have no such requirement.

Warning: Read that explanation one more time. It says:

> If $f: X \to Y$ is a function, then for *any* $x \in X$, we must be able to talk about *the* $y = f(x) \in Y$ that corresponds to x.

Note the emphasis on *any* and *the*. The first tells us that there must not be any values of $x \in X$ at which f is not defined; in other words, the function must be defined at every point in its domain. The second tells us that $f(x)$ must be *uniquely* defined for each $x \in X$; in other words, for every possible value of $x \in X$, there can exist only one value of $y \in Y$ that satisfies the equation $y = f(x)$. □

Let X and Y be sets. Let G be a relation from X to Y that satisfies the following property:[5]

$$\text{for any } x \in X, \text{ there is exactly one } y \in Y \text{ such that } (x, y) \in G. \tag{5.1}$$

Then we will define a function $f: X \to Y$ by the rule

$$y = f(x) \qquad \text{if and only if} \qquad (x, y) \in G. \tag{5.2}$$

Remark. The rule (5.2) defines a function precisely when (5.1) holds. To see why, let's see what could go wrong when trying to use (5.2) to define a function:

- There might be an $x \in X$ such that *no* element $y \in Y$ exists such that $(x, y) \in R$. In this case, there is no y-value that would allow us to define $f(x)$. Put yet another way, this x would be an element in the domain of f at which we could not define f.

- There might be an $x \in X$ such that *more than one* element $y \in Y$ exists such that $(x, y) \in R$. In this case, we wouldn't know which of these y-values to choose as $f(x)$. Note that this would lead to a function with several values at some point.

Since neither of these can go wrong when (5.1) holds, the rule (5.2) defines a function. □

Let's look at some examples of purported functions.

[5] Why did we choose the letter G to name the relation? Simply because "g" is the first letter in the word "graph."

Example 5.6: The following examples are closely related to each other. If you read them carefully, you'll have a better idea of which situations yield functions, and which do not.

(a) Let $G = \{(1, b), (1, c), (2, c), (4, e)\}$. Does G describe a function with domain $\{1, 2, 3, 4\}$ and codomain $\{a, b, c, d, e\}$?

(b) Let $G = \{(1, c), (2, c), (4, e)\}$. Does G describe a function with domain $\{1, 2, 3, 4\}$ and codomain $\{a, b, c, d, e\}$?

(c) Let $G = \{(1, c), (2, c), (3, a), (4, e), (5, e)\}$. Does G describe a function with domain $\{1, 2, 3, 4\}$ and codomain $\{a, b, c, d, e\}$?

(d) Let $G = \{(1, c), (2, c), (3, a), (4, e)\}$. Does G describe a function with domain $\{1, 2, 3, 4\}$ and codomain $\{a, b, c, d, e\}$?

(e) Let $G = \{(1, c), (2, c), (3, a), (4, b)\}$. Does G describe a function with domain $\{1, 2, 3, 4\}$ and codomain $\{a, b, c\}$?

Solution: Pay attention to the similarities and differences here!

(a) No. Both $(1, b)$ and $(1, c)$ are elements of G.

(b) No. Although 3 is an element of the domain, there is no pair of the form (3, something) in G.

(c) No. There's an element $(5, e) \in G$, but 5 is not an element of the domain.

(d) Yes, G is the graph of a function $f : \{1, 2, 3, 4\} \to \{a, b, c, d, e, \}$, defined by the table

x	1	2	3	4
$f(x)$	c	c	a	e

Do not be misled by the fact that the codomain contains b and e, which do not appear in the second position of any element of G. Also, do not be misled by the fact that c appears twice, being paired with both 1 and 2. Both of these are perfectly permissible.

(e) Yes, G is the graph of a function $g : \{1, 2, 3, 4\} \to \{a, b, c\}$, defined by the table

t	1	2	3	4
$g(t)$	c	c	a	b

Once again, do not be misled by the fact that c appears twice, being paired with both 1 and 2. □

Let's look more closely at parts 4 and 5 in Example 5.6. seem fairly similar. But there is an important difference between them:

- In part 4, there are elements in the codomain that are not values of f. In other words, the function f *does not* "hit" all the codomain elements.

- In part 5, every element in the codomain of g is a value of g. In other words, the function g *does* hit all the codomain elements.

Another way of looking at this would to be to say that the codomain of f is too big, but the codomain of g is exactly the right size. Let's make this idea more precise.

Let $f: X \to Y$ be a function. The *range* of f, which we denote "Range(f)," is the set of all the values that f can take, i.e.,

$$\text{Range}(f) = \{ f(x) : x \in X \}.$$

It is always true that $\text{Range}(f) \subseteq Y$, i.e., the range is always a subset of the codomain. Sometimes the range will be a proper subset of its codomain; sometimes the range will equal the whole codomain.

Note that if the range of a function is smaller than the codomain, then we might think about ignoring the excess members of the codomain, i.e., replacing the codomain with the range. In terms of practical implementation (say, on an interactive website), this means that we might be able to improve the efficiency of the implementation, which may make for a faster response time; this may improve how visitors will feel about our website, and keep them coming back as return customers.

Example 5.7: For each of the following functions, determine their ranges and state whether the range is a proper subset of the codomain.

(a) The function $g: \{1, 2, 3, 4\} \to \{1, 2, 3\}$ defined by

t	1	2	3	4
$g(t)$	3	1	2	1

(b) The function $h: \{1, 2, 3, 4\} \to \{1, 2, 3\}$ defined by

t	1	2	3	4
$h(t)$	3	1	3	1

Solution:

(a) We see that Range(g) $= \{1, 2, 3\}$. So g is a function whose range coincides with its codomain.

(b) We see that Range(h) $= \{1, 3\}$. Note that Range(h) $\neq \{1, 2, 3\}$, since 2 belongs to the codomain of h but $h(t) \neq 2$ for all $t \in \{1, 2, 3, 4\}$. So h is a function whose range is a proper subset of its codomain. □

Example 5.8: [SN] Look at Table A.1 in the Appendix, which gives basic information for our sample social network. Any pair of distinct columns (e.g., (Name, City), (Hometown, Status), (Sex, Hometown)) will define a binary relation from the attribute named by the first column to that named by the second column. Which of these relations are functions?

Solution: Since we require the entries in the *Name* column to be distinct, all the relations whose codomain is *Name* will be functions, i.e., Name $\rightarrow$ City, Name $\rightarrow$ Hometown, Name $\rightarrow$ Sex, Name $\rightarrow$ Birthday, and Name $\rightarrow$ Status are all functions. Generally speaking, one would not expect any of the other pairs to yield a function. For example, consider (Birthday, City); a functional relationship would mean that two people with the same birthday would need to live in separate cities.

However, if we were to only consider the actual data in the table, rather than the usual relationships between the attributes named by the table columns, we get a surprise. It turns out that Birthday $\rightarrow$ City, Birthday $\rightarrow$ Hometown, Birthday $\rightarrow$ Sex, and Birthday $\rightarrow$ Status happen to be functional relationships, simply because the data in the table just happened to make things work out that way; had the data been chosen differently, these might not have been functional relationships. □

Based on Example 5.8, we can distinguish between

- *essential* functional relationships, in which the functional relationship is inherently part of the underlying rule that defines the relationship, and

- *accidental* functional relationships, in which there's no a priori[6] functional relationship between the domain and codomain, but the data in the relationship table just happens to display a functional nature.

[6] *A priori* refers to knowledge that is derived from self-evident propositions rather than from observation or experience.

If there's no underlying rule that defines a binary relation (i.e., if we're simply given a set of ordered pairs), then we generally can't tell the difference between essential and accidental functional relationships.

5.3 Properties of Functions

In Section 5.3, we described useful properties (reflexivity, irreflexivity, symmetry, antisymmetry, reflexivity) that relations might satisfy. You probably won't be surprised that there are useful properties that functions sometimes satisfy, as well.

These properties are as follows:

- A function $f: S \to T$ is *injective* if $f(x) = f(y)$ always implies that $x = y$, for any $x, y \in S$.

  It's sometimes more convenient to work with the contrapositive of this definition (which, as we know from page 101, is equivalent to the definition itself): A function $f: S \to T$ is injective if for any two $x, y \in S$ with $x \neq y$, we always have $f(x) \neq f(y)$.

- A function $f: S \to T$ is *surjective* if for any $t \in T$, there exists some $s \in S$ such that $t = f(s)$. An equivalent formulation would be that $\text{Range}(f) = T$.

- A function $f: S \to T$ is *bijective* if f is both injective and surjective.

Here is some alternative terminology:

- We sometimes replace the adjectives "injective," "surjective," and "bijective" with nouns. Thus we might say "f is an injection," "f is a surjection," or "f is a bijection," instead of saying "f is injective," "f is surjective," or "f is bijective."

- Some people like to use simpler language. So they'll say that "f is *one-to-one*" or that "f maps S to T in a *one-to-one* manner," rather than saying "$f: S \to T$ is injective." Similarly, they will say "f maps S *onto* T," rather than saying "$f: S \to T$ is surjective." Finally, you will sometimes say that $f: S \to T$ is a *one-to-one correspondence*, rather than saying that "$f: S \to T$ is bijective."

Warning: Unfortunately, you will sometimes hear the linguistic barbarism, "The function f is onto." The word "onto" is a preposition, and not an adjective. Please do not commit this crime against the English language. (The authors thank you in advance.) □

What does all this mean? Think of a function $f: S \to T$ as labeling each point in S with a value in T.

- For f to be injective, no two distinct points in S can have the same label.
- For f to be surjective, every point in T must have *at least* one label.
- For f to be bijective, every point in T must have *exactly* one label.

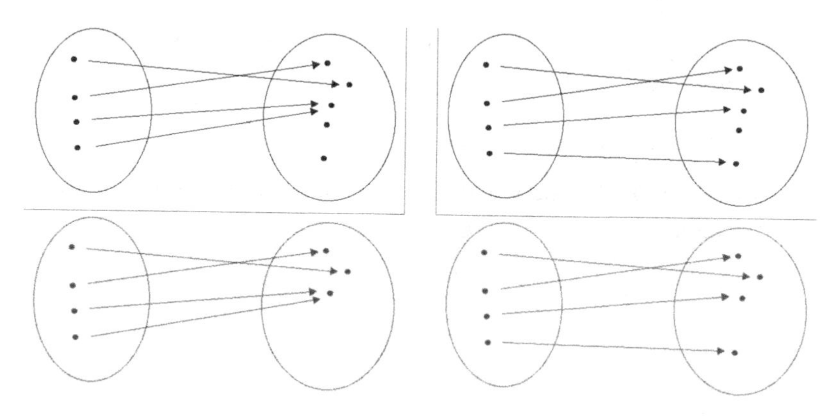

Figure 5.4: Properties of functions

Figure 5.4 gives another way of looking at these properties:

- The function in the upper-left-hand corner of the figure is neither injective nor surjective. Note that there is a point in the codomain that is hit more than once, as well as a point in the codomain that is not hit at all.
- The function in the upper-right-hand corner of the figure is injective, but not surjective. Note that no point in the codomain is hit more than once, but there is a point in the codomain that is not hit at all.
- The function in the lower-left-hand corner of the figure is surjective, but not injective. Note that every point in the codomain is hit at least once, but there is a point that is hit more than once.

- The function in the lower-right-hand corner is bijective. Every point in the codomain is hit exactly once.

Some people have found the following analogy to be a useful way of visualizing the concepts of injectivity, surjectivity, and bijectivity. Let C be a can of paint and let F be a floor. Suppose that we transfer the paint to the floor (either by painting the floor or by spilling the can of paint). Define a function $p\colon C \to F$ by the following rule: for each drop of paint d, let $p(d)$ be the spot on the floor where the paint drop d finally winds up.

- If no spot on floor winds up with more than one drop of paint, then p is injective.
- If the entire floor gets covered with paint, then p is surjective.
- If every spot on entire floor gets covered with exactly one drop of paint, then p is bijective.

Let's look at a few examples:

Example 5.9: Determine whether each of the following functions is injective, bijective, and/or surjective.

(a) The function $h\colon \{1, 2, 3, 4\} \to \{1, 2, 3\}$ defined by

t	1	2	3	4
$h(t)$	3	1	2	2

(b) The function $f\colon \{1, 2, 3\} \to \{1, 2, 3, 4\}$ defined by

s	1	2	3
$f(s)$	3	2	1

(c) The function $s\colon \{1, 2, 3, 4\} \to \{\clubsuit, \diamondsuit, \heartsuit, \spadesuit\}$ defined by

τ	1	2	3	4
$s(\tau)$	$\spadesuit$	$\heartsuit$	$\diamondsuit$	$\clubsuit$

Solution:

(a) The function h is not injective, since $h(3) = h(4) = 2$. However, h is surjective. Of course, h is not bijective, since it is not injective.

(b) The function f is injective. However, f is not surjective, since $f(t)$ never equals 4. Of course, f is not bijective, since it is not surjective.

(c) The function s is both injective and surjective. So, s is bijective. □

Example 5.10: [SN] In Example 5.8, we have five non-accidental functions. Which of them are injective? surjective? bijective?

Solution: None of these functions are injective, since Kip and Kyle share the same city, hometown, sex, birthday, and status. For surjectivity:

- The sex function is surjective (Alex is female, Erik is male), as is the status function (Alex is single, Anna is undeclared, and Ellen is married).
- The city, hometown, and birthday functions are not surjective; for instance, nobody's city or hometown is Trenton (NJ), and nobody was born on 12/22/1950.

Since none of these functions are injective, none of them can be bijective functions. □

It turns out that there are some simple tests that you can use that will totally rule out the possibility of an injection, surjection, or bijection between sets, based on the cardinalities (sizes) of those sets:

Fact. *Let A and B be finite sets.*

1. *If $|A| < |B|$, then there can be no surjection from A to B.*
2. *If $|A| > |B|$, then there can be no injection from A to B.*
3. *If $|A| \neq |B|$, then there can be no bijection from A to B.* □

Item (2) above is sometimes called the *pigeonhole principle*: If m pigeons are trying to fly into n pigeonholes and $m > n$, then some pigeonhole will wind up with more than one pigeon, see Figure 5.5.

Example 5.11: Show that there are two people in New York, N.Y. having the same number of friends.

Solution: For the sake of argument, let's assume that nobody in New York City has more than (say) 10,000 friends. (We're ruling out people who say that everybody is their friend.) According to the 2010 U.S. Census, there are 8,175,133 people living in Manhattan. We claim that there are at least two people in Manhattan having the same number of friends. To

Figure 5.5: The pigeonhole principle (with $m = 10$ and $n = 9$)

see this, let M denote the set of people living in Manhattan, and let $S = \{0, 1, 2, \ldots, 10{,}000\}$. Define a function $f : M \to S$ by letting $f(m)$ be the number of friends that the Manhattanite $m \in M$ has. Since $|S| < |M|$, the function f cannot be an injection. Hence there are distinct $x, y \in M$ such that $f(x) = f(y)$, i.e., the Manhattanites x and y have the same number of friends. $\square$

5.4 Function Composition

Suppose that we wanted to write a particularly complicated piece of software. For example, deciding that you were dissatisfied with the quality of existing word processors or web browsers, you might decide to write one yourself. As another example, you might decide that you want to start up your own social network; before doing the necessary publicity that would attract new members to your network, the software that will run the network needs to be in place and ready to roll.

We wouldn't want to write any of these programs as a single monolithic block of code. Such an undertaking would require us to keep track of far too many details at once. In addition, suppose that we wanted a team of programmers to work on this large program; dividing up the responsibilities among the team members would be difficult. For example, all the programmers would need to agree on specific naming conventions that are used within the program. This alone would necessitate a lot of administrative overhead in the form of meetings or memos. Instead, we would design the program as a set of small independent parts (often called *modules*). If we use this idea, then each programmer could work on her own assigned set of modules, without worrying about how her work would affect the other programmers. In other words, we've *decomposed* a complicated task into a set of simpler tasks. For example, software running a social network would have modules for locating potential new friends, modifying profiles, sending messages between friends, and so forth.

Of course, it might be the case that these simpler pieces might themselves be too complicated to do all at once, and so that they might also need to be decomposed into simpler modules. This process of designing by decomposition could continue until we reach a point that the modules are easy to write without further decomposition. Such an approach is called *top-down design*, which may be thought of as "design by decomposition."

Rather than trying to write a complicated computer program, suppose that we were given the similar task of computing a complicated function. Let's try to apply this idea of decomposing a complicated task into a set of subtasks. In other words, let's see if we can write the complicated function using a set of simpler "sub-functions."

Example 5.12: Suppose we were to define $h\colon \mathbb{R} \to \mathbb{R}$ as

$$h(x) = (2x+1)^2 + 32(2x+1) \qquad \forall x \in \mathbb{R}.$$

Note that the value $2x+1$ appears twice in $h(x)$; the first time it appears, it gets squared, whereas the second time it appears, it gets multiplied by 32. So rather than computing $2x+1$ twice (once for each of these appearances), it makes sense to *precompute* this value, saving the precomputed value, say, as y. Then $h(x)$ is simply $y^2 + 32y$.

Let's think about the fact that $h(x) = y^2 + 3y$, where $y = 2x + 1$. We can write $y = f(x)$, where the function $f\colon \mathbb{R} \to \mathbb{R}$ is defined as

$$f(x) = 2x + 1 \qquad \forall x \in \mathbb{R}.$$

Similarly, we can also write $z = g(y)$, where the function $g\colon \mathbb{R} \to \mathbb{R}$ is defined as

$$g(y) = y^2 + 32y \qquad \forall y \in \mathbb{R}.$$

Now in the previous paragraph, we calculated $y = 2x + 1$, which is the same as $y = f(x)$; we then calculated $z = y^2 + 32y$, which is the same as $z = g(y)$. But $z = h(x)$, the only possible new idea here being that we computed it in two stages. So

$$h(x) = z = g(y) = g\big(f(x)\big),$$

the notation $g\big(f(x)\big)$ being read as "g evaluated at $f(x)$." Leaving out the intermediate steps, we see that

$$h(x) = g\big(f(x)\big) \qquad \forall x \in \mathbb{R}. \qquad \square$$

Of course, we could also take the opposite viewpoint to program design. We start by building simple modules, combining them to get modules that solve more complex parts of our problem. We could then continue by combining *these* modules, and so forth, until we eventually arrive at a solution to the whole problem. This program design methodology is called *bottom-up design*, which may be thought of as "design by composition."

What do we learn from all this discussion? On the one hand, we often want to decompose a complicated function into a series of simpler functions. On the other hand, we may also want to go in the opposite direction, creating a complicated function by combining some simple functions together; since the first process is a decomposition, it's natural to refer to this process (which is the opposite of the first) as being "composition."

We now describe this process of functional composition more precisely. Let $f: X \to Y$ and $g: Y \to Z$ be functions. The *composite function* $g \circ f: X \to Z$ (read this as "g composed with f") is defined as

$$(g \circ f)(x) = g\big(f(x)\big) \qquad \forall x \in X.$$

Figure 5.6 illustrates this process.

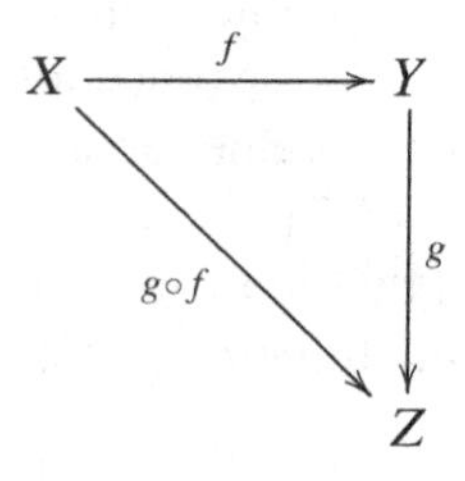

Figure 5.6: Composite functions

Warning: Although we write $g \circ f$ and we read g before f when we say "g composed with f," we first calculate $y = f(x)$ and then $z = g(y)$ when we compute $z = g\big(f(x)\big)$. Sometimes people read $g \circ f$ as "g after f," which places an emphasis on the order in which the functions are applied. □

Remark. Computing such a composite function $g \circ f$ at a point $x \in X$ is a two-step process:

1. Compute $y = f(x)$.

2. Compute $z = g(y)$.

Then $z = (g \circ f)(x)$. Note that *the output of the first function* becomes *the input to the second function.* □

Example 5.13: Define $f : \mathbb{R} \to \mathbb{R}$ by

$$f(x) = 2x \qquad \forall x \in \mathbb{R}$$

and $g : \mathbb{R} \to \mathbb{R}$ by

$$g(x) = x + 1 \qquad \forall x \in \mathbb{R}$$

Compute $(f \circ g)(2)$ and $(g \circ f)(2)$.

Solution: We have

$$(f \circ g)(2) = f(g(2)) = f(2 + 1) = f(3) = 2 \times 3 = 6;$$

alternatively, we could calculate this as

$$(f \circ g)(2) = f(g(2)) = 2 \times g(2) = 2 \times (2 + 1) = 6.$$

For second part, we have

$$(g \circ f)(2) = g(f(2)) = f(2) + 1 = 2 \times 2 + 1 = 5;$$

alternatively, we could calculate this as

$$(g \circ f)(2) = g(f(2)) = g(2 \times 2) = g(4) = 4 + 5.$$

Note that $(g \circ f)(2) \neq (f \circ g)(2)$, i.e., function composition is not commutative. □

Example 5.14: Let P be the set of all people. Define functions $f : P \to P$ and $m : P \to P$ by the rules

$$f(p) = \text{the (birth) father of } p \qquad \forall p \in P$$

and

$$m(p) = \text{the (birth) mother of } p \qquad \forall p \in P.$$

What is $m \circ m$? $m \circ f$? $f \circ m$? $f \circ f$?

Solution: For any person $p \in P$, we find that $(m \circ m)(p) = m\big(m(p)\big)$, which is the mother of the mother of p, i.e., the maternal grandmother of p. Similarly, we find that

$$m \circ f = \text{paternal grandmother,}$$
$$f \circ m = \text{maternal grandfather,}$$

and

$$f \circ f = \text{paternal grandfather}.$$

Note that $f \circ m \neq m \circ f$, i.e., composing the functions f and m is not commutative. □

Remark. It would be hard to extend Example 5.14 to include siblings, not to mention family relationships that involve siblings (such as aunt, uncle, or cousin). That's because a given person can have any number of siblings (and aunts, uncles, or cousins), including none whatsoever. These other family relationships are examples of non-functional relationships.[7] □

In the Remark on page 168, we saw that when composing two functions, the output of the first function becomes the input of the second function. There's no reason to restrict this to composing two functions; we can compose as many functions as we like, the only restriction being that the range of a particular function must be a subset of the domain of the next function in the "chain." For example, suppose that $f: W \to X$, $g: X \to Y$, and $h: Y \to Z$ are functions. Then we can define $h \circ g \circ f: W \to Z$ by the rule

$$(h \circ g \circ f)(w) = h\Big(g\big(f(w)\big)\Big) \qquad \forall\, w \in W.$$

Note that $\circ$ is a binary operation. Strictly speaking, one should write either $h \circ (g \circ f)$ or $(h \circ g) \circ f$, depending on which composition is done first. It is not very hard to show that function composition is associative, i.e., that

$$h \circ (g \circ f) = (h \circ g) \circ f,$$

and so the parentheses are unnecessary.

Warning: Keep in mind that function composition is associative, but not commutative. □

Example 5.15: Define functions $f, g, h: \mathbb{R} \to \mathbb{R}$ by

$$f(x) = x + 1 \qquad \forall\, x \in \mathbb{R},$$
$$g(x) = 4x \qquad \forall\, x \in \mathbb{R},$$
$$h(x) = \tfrac{1}{2}x - 1 \qquad \forall\, x \in \mathbb{R}.$$

[7] See footnote on page 150.

What is $h \circ (g \circ f): \mathbb{R} \to \mathbb{R}$? What is $(h \circ g) \circ f: \mathbb{R} \to \mathbb{R}$?

Solution: We have

$$g\big(f(x)\big) = g(x+1) = 4(x+1) = 4x+4 \qquad \forall\, x \in \mathbb{R},$$

and so

$$\begin{aligned}(h \circ g \circ f)(x) &= h\Big(g\big(f(x)\big)\Big) = \tfrac{1}{2}g\big(f(x)\big) - 1 \\ &= \tfrac{1}{2}(4x+4) - 1 \qquad\qquad \forall\, x \in \mathbb{R}. \\ &= 2x + 2 - 1 = 2x + 1\end{aligned}$$

On the other hand,

$$(h \circ g)(y) = h(4y) = \tfrac{1}{2}(4y) - 1 = 2y - 1 \qquad \forall\, y \in \mathbb{R},$$

and so letting $y = f(x)$, we have

$$\begin{aligned}\big((h \circ g) \circ f\big)(x) &= (h \circ g)\big(f(x)\big) = 2f(x) - 1 \\ &= 2(x+1) - 1 = 2x + 2 - 1 = 2x + 1\end{aligned} \qquad \forall\, x \in \mathbb{R}.$$

Note that $h \circ (g \circ f) = (h \circ g) \circ f$. □

In the previous example, the order in which we computed the composite function did not matter; we could either compute $g \circ f$, and then apply h to the result, or we could figure out what $h \circ g$ should be and then compose this new function with f. This is true in general:

Fact. *Let $f: W \to X$, $g: X \to Y$, and $h: Y \to Z$ be functions with the given domains and codomains. Then*

$$h \circ (g \circ f) = (h \circ g) \circ f,$$

i.e., function composition is associative. □

In Exercise 5.9.9, we let you try your hand at showing this for yourself.

5.5 Identity and Inverse Functions

We'll now discuss a function that at first sight may seem to be totally useless; we'll come to understand its importance later on. A given set's *identity function* is the function that does absolutely nothing whatsoever. More precisely, let S be any set. The *identity function* id_S *on* S is defined by the rule

$$\mathrm{id}_S(x) = x \qquad \forall x \in S.$$

Note that if $f\colon X \to Y$, then

$$f \circ \mathrm{id}_X = f = \mathrm{id}_Y \circ f.$$

Why do we call these functions "identity functions?" Think about the simple fact

$$a \times 1 = a = 1 \times a \qquad \forall a \in \mathbb{R}.$$

Comparing this to the previous equation, we see how identity functions are analogous to the multiplicative identity (i.e., the number 1), and hence the name is well-chosen.

Example 5.16: Let $V = \{\mathtt{a}, \mathtt{e}, \mathtt{u}, \mathtt{o}, \mathtt{u}\}$ be the set of all vowels in the English alphabet, and let C the set of all consonants. Describe the identity functions $\mathrm{id}_V\colon V \to V$ and $\mathrm{id}_C\colon C \to C$.

Solution: The identity function id_V is given by the table

x	a	e	i	o	u
$\mathrm{id}_V(x)$	a	e	i	o	u

The identity function id_C is given analogously. □

One's first impression might well be that such a simple "do-nothing" function is absolutely worthless. As often happens with first impressions, nothing could be further from the truth. But before we convince you of this fact, we need to introduce one more concept, namely, invertible functions.

We say that a function $f\colon X \to Y$ is *invertible* if there exists a function $f^{-1}\colon Y \to X$ such that

$$f^{-1} \circ f = \mathrm{id}_X \qquad \text{and} \qquad f \circ f^{-1} = \mathrm{id}_Y.$$

In other words, $f\colon X \to Y$ is invertible if and only if there exists a function $f^{-1}\colon Y \to X$ such that

$$\begin{aligned} f^{-1}\big(f(x)\big) &= x \qquad \forall x \in X \\ f\big(f^{-1}(y)\big) &= y \qquad \forall y \in Y. \end{aligned}$$

We say that f^{-1} is the functional *inverse* of f. In other words, a function is invertible if there is another function that "undoes" the action of the first function.

Warning: Unfortunately, -1 in a superscript position has two uses:

- If x is a number or a variable, then x^{-1} denotes the multiplicative inverse of x, i.e., $x^{-1} = 1/x$.
- If f is a function, then f^{-1} denotes the functional inverse of f, as defined above.

Each is an "inverse" of some kind, but the two uses are incompatible, i.e., you can't write

$$f^{-1}(x) = \frac{1}{f(x)}.$$

Be careful to pay attention the context in which the superscript -1 appears! □

Let's look at a couple of examples of functions and their inverses:

Example 5.17: Define the function $g\colon \mathbb{Z} \to \mathbb{Z}$ by the rule

$$g(x) = x - 7 \qquad \forall\, x \in \mathbb{Z}.$$

Show that $g^{-1}\colon \mathbb{Z} \to \mathbb{Z}$ is given by

$$g^{-1}(y) = y + 7 \qquad \forall\, y \in \mathbb{Z}.$$

Solution: We have

$$\begin{aligned} (g \circ g^{-1})(y) &= g\big(g^{-1}(y)\big) = g(y+7) \\ &= (y+7) - 7 = y \end{aligned} \qquad \forall\, y \in \mathbb{Z}$$

and

$$\begin{aligned} (g^{-1} \circ g)(x) &= g^{-1}\big(g(x)\big) = g^{-1}(x-7) \\ &= (x-7) + 7 = x \end{aligned} \qquad \forall\, x \in \mathbb{Z}.$$

So g^{-1} is the functional inverse of g, as claimed. □

Example 5.18: Define the function $h\colon \mathbb{R} \to \mathbb{R}$ by

$$h(x) = \tfrac{1}{2}x \qquad \forall\, x \in \mathbb{R}.$$

Show that $h^{-1}\colon \mathbb{R} \to \mathbb{R}$ is given by

$$h^{-1}(y) = 2y \qquad \forall\, y \in \mathbb{R}.$$

Solution: We have

$$(h \circ h^{-1})(y) = h\big(h^{-1}(y)\big) = \tfrac{1}{2}h^{-1}(y) = \tfrac{1}{2} \times 2y = y \qquad \forall\, y \in \mathbb{R}$$

and

$$(h^{-1} \circ h)(x) = h^{-1}\big(h(x)\big) = 2h(x) = 2 \times \tfrac{1}{2}x = x \qquad \forall\, y \in \mathbb{R}.$$

So h^{-1} is the functional inverse of h, as claimed. □

But not all functions are invertible. In fact, the distinction between an invertible and a non-invertible function may be quite subtle, as we now show:

Example 5.19: Define the function $m: \mathbb{Q} \to \mathbb{Q}$ by

$$m(x) = 2x \qquad \forall\, x \in \mathbb{Q}.$$

Is m invertible? If so, what is its inverse function?

Solution: We need to solve the equation

$$y = m(x) = 2x$$

for x in terms of y. Using simple algebra, we have

$$y = 2x \iff x = \tfrac{1}{2}y.$$

Clearly, $y \in \mathbb{Q}$ implies that $x = \frac{1}{2}y \in \mathbb{Q}$. Moreover, the $\iff$ tells us that for any $y \in \mathbb{Q}$, there is exactly one $x \in \mathbb{Q}$ for which $y = m(x)$, namely, $x = \frac{1}{2}y$. Thus m is invertible, and $m^{-1}: \mathbb{Q} \to \mathbb{Q}$ is given by the rule

$$m^{-1}(y) = \tfrac{1}{2}y \qquad \forall\, y \in \mathbb{Q}. \quad \square$$

Example 5.20: Let's make a slight change to the function m in Example 5.19. Define a new function $\widetilde{m}: \mathbb{N} \to \mathbb{N}$ by

$$\widetilde{m}(x) = 2x \qquad \forall\, x \in \mathbb{N}.$$

Is $\widetilde{m}$ invertible? If so, give its inverse function.

Solution: As in Example 5.19, we need to solve the equation

$$y = \widetilde{m}(x) = 2x$$

for x in terms of y. Again, we have

$$y = 2x \iff x = \tfrac{1}{2}y.$$

However $y \in \mathbb{N}$ does *not* imply that the resulting x is a natural number. For example, if $y = 1$, then $x = \frac{1}{2}y = \frac{1}{2}$ is not a natural number. So $\widetilde{m}$ is *not* invertible. $\square$

So how can we determine whether a given function is invertible?

Fact. *The function $f: X \to Y$ is invertible if and only if f is a bijection.*

Remark. Why is this true? Since $x = f^{-1}(f(x))$ always holds, we can substitute $f(x) = y$ to see we must have

$$x = f^{-1}(y) \iff y = f(x).$$

This looks as if it defines a function f^{-1}. But is f^{-1} really a *function* from Y to X, or is it merely a relation? For f^{-1} to be a function means that for any $y \in Y$, there must exist a unique $x \in X$ such that $x = f^{-1}(y)$, i.e., such that $y = f(x)$. Note that the "uniqueness" is equivalent to requiring that f be an injection; indeed, if $y = f(x)$ and also $y = f(x')$, we wouldn't know whether we should use x or x' as the value of $f^{-1}(y)$. Likewise, the "existence" part can happen if and only if for any $y \in Y$, there exists some $x \in X$ such that $f(x) = y$; this is equivalent to asking that f be a surjection. □

Let's look at some examples.

Example 5.21: Let $f: \{1, 2, 3, \ldots, 999\} \to \{0, 1, 2, 3, \ldots, 999\}$. Note that we have not given a rule that defines f; we are letting f be *any* function with this domain and codomain. Is it possible for f to be invertible?

Solution: Since the domain is smaller than the domain, the Pigeonhole Principle tells us that no such function can be an injection. Hence no such function can be invertible. □

Example 5.22: Define a function $q: \{1, 2, 3, 4\} \to \{\clubsuit, \diamondsuit, \heartsuit, \spadesuit\}$ by

τ	1	2	3	4
$q(\tau)$	$\spadesuit$	$\heartsuit$	$\diamondsuit$	$\clubsuit$

Is q invertible?

Solution: It's easy to see that q is a bijection, and so q is invertible. For that matter, the table

s	$\clubsuit$	$\diamondsuit$	$\heartsuit$	$\spadesuit$
$q^{-1}(s)$	4	3	2	1

explicitly gives us the inverse of q. □

The next example shows that the domain and codomain are critically important in determining whether a given function is invertible.

Example 5.23: Which of the following functions are invertible? If the function is invertible, give its inverse.

(a) Let $f\colon \mathbb{R} \to \mathbb{R}$ be defined by

$$f(x) = x^2 \qquad \forall x \in \mathbb{R}.$$

(b) Let $g\colon \mathbb{R} \to \mathbb{R}^{\geq 0}$ be defined by

$$g(x) = x^2 \qquad \forall x \in \mathbb{R}.$$

(c) Let $h\colon \mathbb{R}^{\geq 0} \to \mathbb{R}^{\geq 0}$ be defined by

$$h(x) = x^2 \qquad \forall x \in \mathbb{R}.$$

Solution:

(a) The function f is not a surjection, since (e.g.) there is no $x \in \mathbb{R}$ such that $f(x) = -1$. So f is not invertible.

(b) The function g is not an injection, since $g(1) = 1$ and $g(-1) = 1$. So g is not invertible.

(c) This one works.

  (a) We claim that the function h is injective. Indeed,

$$\begin{aligned} h(x) = h(\tilde{x}) &\iff x^2 = \tilde{x}^2 \\ &\iff x^2 - \tilde{x}^2 = 0 \\ &\iff (x - \tilde{x})(x + \tilde{x}) = 0 \\ &\iff x = \tilde{x} \quad \text{or} \quad x = -\tilde{x}. \end{aligned}$$

  Thus $h(x) = h(\tilde{x})$ if and only $x = \tilde{x}$ or $x = -\tilde{x}$. Since we are dealing with non-negative numbers, we cannot have $x = -\tilde{x}$ unless both x and $\tilde{x}$ are 0, which would imply $x = \tilde{x}$. So $h(x) = h(\tilde{x})$ if and only $x = \tilde{x}$. Thus h is injective, as claimed.

  (b) We claim that the function h is surjective. Indeed, let $y \in \mathbb{R}^{\geq 0}$. We need to find $x \in \mathbb{R}^{\geq 0}$ such that $h(x) = y$, i.e., $x^2 = y$. Let's choose $x = \sqrt{y}$. Then

$$h(x) = h\left(\sqrt{y}\right) = \left(\sqrt{y}\right)^2 = y,$$

  as required.

Thus f is bijective, and hence f is invertible. In fact, its inverse function $f^{-1}\colon \mathbb{R}^{\geq 0} \to \mathbb{R}^{\geq 0}$ is defined by

$$f^{-1}(y) = \sqrt{y} \qquad \forall\, y \in \mathbb{R}^{\geq 0}. \qquad \square$$

So now we know how to determine whether or not a function is invertible. Next, suppose that you have been given an invertible function $f\colon X \to Y$. Here's how you could find its inverse function $f^{-1}\colon Y \to X$, based on the fact that

$$x = f^{-1}(y) \qquad \text{if and only if} \qquad y = f(x). \tag{5.3}$$

The steps are as follows:

1. Write down the equation $y = f(x)$ or (equivalently) $f(x) = y$.
2. Solve the equation $f(x) = y$ for x in terms of y. Note that this may be easy or difficult (or impossible!), depending on the function f. In principle, you need to verify the following:
   - There must be exactly one solution that gives x in terms of y (otherwise the inverse "function" won't be a true function).
   - Since the inverse function goes from Y to X, you need to check that if $y \in Y$, then the resulting x value is an element of X.

You'll now have something of the form

$$x = \text{some expression involving } y. \tag{5.4}$$

Looking at (5.4), we see that the expression involving y on the right hand side of (5.4) is precisely $f^{-1}(y)$.

Example 5.24: Define the function $q\colon \mathbb{Z} \to \mathbb{Z}$ by the rule

$$q(x) = x + 1 \qquad \forall x \in \mathbb{Z}.$$

Is q invertible? If so, what is its inverse function?

Solution: We need to solve the equation

$$y = q(x) = x + 1$$

for x in terms of y. Using simple algebra, we have

$$y = x + 1 \qquad \Longleftrightarrow \qquad x = y - 1.$$

Clearly $y \in \mathbb{Z}$ implies that $x = y - 1 \in \mathbb{Z}$. Moreover, the $\iff$ tells us that for any $y \in \mathbb{Z}$, there is exactly one $x \in \mathbb{Z}$ such that $y = q(x)$, namely, $x = y - 1$. So q is invertible, and $q^{-1}\colon \mathbb{Z} \to \mathbb{Z}$ is given by the rule

$$q^{-1}(y) = y - 1 \qquad \forall\, y \in \mathbb{Z}. \qquad \square$$

Note that we don't always need to use x and y to denote variables in the domain and the codomain:

Example 5.25: Define the function $g\colon \mathbb{R} \to \mathbb{R}$ by the rule

$$g(s) = 4s - 3 \qquad \forall\, s \in \mathbb{R}.$$

Is g invertible? If so, what is its inverse function?

Solution: We need to solve the equation

$$t = g(s) = 4s - 3$$

for s in terms of t, then $s = g^{-1}(t)$. Using simple algebra, we have

$$t = 4s - 3 \iff t + 3 = 4s \iff s = \tfrac{1}{4}(t + 3).$$

It is easy to see that $t \in \mathbb{R} \implies s \in \mathbb{R}$. Moreover, for any $t \in \mathbb{R}$, the $\iff$ tells us that there is exactly one $s \in \mathbb{R}$ such that $t = g(s)$, namely, $s = \frac{1}{4}(t + 3)$. Thus g is invertible, and $g^{-1}\colon \mathbb{R} \to \mathbb{R}$ is given by the rule

$$g^{-1}(t) = \tfrac{1}{4}(t + 3) \qquad \forall\, t \in \mathbb{R}. \quad \square$$

Now let's talk about inverting composite functions. Recall that in Section 5.4, we introduced function composition as a tool to help us manage the complexity of building large, complicated systems. We try to break what might be an overwhelming task into a series of less daunting subtasks. This same idea can apply to the inversion of complicated functions. Rather than trying to invert a difficult function all at once, let's break it up as the composition of simpler functions. Suppose that we can invert these simpler functions; how can we determine the inverse of the composite function?

To keep things simple, let's look at trying to invert a function that is the composite of two simpler functions. To be specific, suppose that $f\colon X \to Y$ and $g\colon Y \to Z$ are invertible functions, and our task is to find the inverse of $h = g \circ f$. Going back and looking at Figure 5.6 on page 168, we see that $h\colon X \to Z$. Now recall that the domain of a function is the codomain of its inverse, and vice-versa. Thus we know that

$$f^{-1}\colon Y \to X \qquad \text{and} \qquad g^{-1}\colon Z \to Y,$$

and that

$$h^{-1}\colon Z \to X \qquad \text{(provided that } h \text{ is invertible to begin with!).}$$

Figure 5.7 neatly summarizes our situation. From the viewpoint of which functions can be composed with each other, it seems that if $h = g \circ f$ is invertible, then we should have $h^{-1} = f^{-1} \circ g^{-1}$; there's simply no other way to put the functions together. So let's check to see whether our

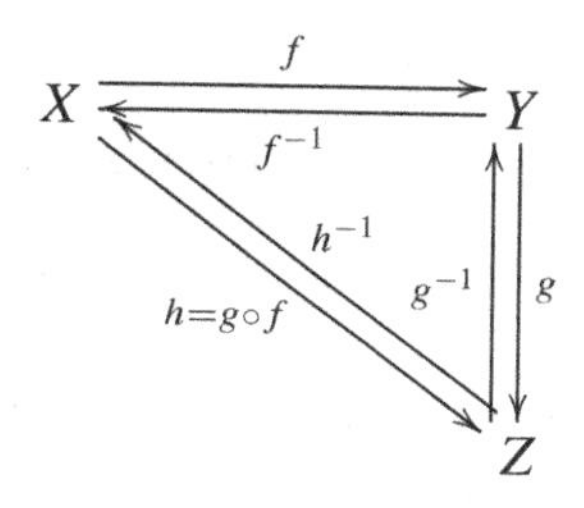

Figure 5.7: Inverting composite functions

conjecture that $h = g \circ f$ is invertible, with $h^{-1} = f^{-1} \circ g^{-1}$. From the definition of "inverse function," we see that we need to prove

$$(f^{-1} \circ g^{-1}) \circ h = \mathrm{id}_X \qquad \text{and} \qquad h \circ (f^{-1} \circ g^{-1}) = \mathrm{id}_Z \tag{5.5}$$

Let's show that the left-hand equation in (5.5) is true. Recalling that $h = g \circ f$ and using the fact the function composition is associative, we see that

$$\begin{aligned}(f^{-1} \circ g^{-1}) \circ h &= (f^{-1} \circ g^{-1}) \circ (g \circ f) = f^{-1} \circ (g^{-1} \circ g) \circ f \\ &= f^{-1} \circ \mathrm{id}_Y \circ f = f^{-1} \circ f = \mathrm{id}_X,\end{aligned}$$

as required. The proof of the right-hand equation in (5.5) is similar.

We have just established the following:

Fact. *Let $f\colon X \to Y$ and $g\colon Y \to Z$ be invertible functions. Then $g \circ f\colon X \to X$ is invertible, with*

$$(g \circ f)^{-1} = f^{-1} \circ g^{-1}. \qquad \square$$

There was nothing special about only using two functions (f and g) here; this was only done for simplicity's sake. This idea extends to as many composable invertible functions as you like: the inverse of a composite of several functions is the composition of the inverses of the individual functions, but in reverse order.

Example 5.26: Define $h\colon \mathbb{R} \to \mathbb{R}$ by

$$h(x) = 2x - 7 \qquad \forall x \in \mathbb{R}.$$

Determine the function $h^{-1}\colon \mathbb{R} \to \mathbb{R}$.

Solution: Note that for any $x \in \mathbb{R}$, the computation of $h(x) = 2x - 7$ can be broken into two steps: doubling x, and then adding 7 to the result. This means that $h = g \circ f$, where $f\colon \mathbb{R} \to \mathbb{R}$ is the "doubling" function given by

$$f(x) = 2x \qquad \forall x \in \mathbb{R}$$

and $g\colon \mathbb{R} \to \mathbb{R}$ is the "add seven" function given by

$$g(y) = y + 7 \qquad \forall y \in \mathbb{R}.$$

It should be intuitively clear that $f^{-1}\colon \mathbb{R} \to \mathbb{R}$ is given by

$$f^{-1}(y) = \tfrac{1}{2}y \qquad \forall y \in \mathbb{R}$$

(i.e., the opposite of doubling a number is simply dividing a number by 2) and that $g^{-1}\colon \mathbb{R} \to \mathbb{R}$ is given by

$$g^{-1}(z) = z + 7 \qquad \forall z \in \mathbb{R}$$

(i.e., the opposite of adding 7 to a number is subtracting 7 from a number); however, you should feel free to use the definition of an inverse function to formally verify that the inverses of f and g as claimed. Since $h = g \circ f$, it follows that $h^{-1} = f^{-1} \circ g^{-1}$, so that

$$h^{-1}(z) = f^{-1}\big(g^{-1}(z)\big) = f^{-1}(z + 7) = \tfrac{1}{2}(z + 7) \qquad \forall z \in \mathbb{R}. \qquad \square$$

5.6 An Application: Cryptography

As promised, we'll now show some examples that should convince you that identity functions and inverse functions are vitally important.

People often need to send and receive confidential messages, secure in the knowledge that these messages are safe from prying eyes. For example:

- A leader may need to send messages to his subordinates or confidantes.
- A diplomat may need to receive instructions from her government.

- A military officer may need to send battle plans to his troops.
- Somebody visiting an e-business website will need to submit the details of her order, along with some method of payment, such as a credit card.

Should these messages be successfully intercepted, the consequences could be merely annoying, potentially embarrassing, or even disastrous.

Cryptography deals with the problem of hiding information from people who shouldn't see it.[8] The main idea is to replace the original message (called the *cleartext*) with an encrypted message (called a *ciphertext*). The ciphertext is then transmitted to the intended recipient, who can then decrypt the ciphertext, thereby getting the original cleartext message.

We shall see that functional inverses play a key role in computational cryptography.

5.6.1 Caesar Rotation

One of the earliest known examples of cryptography is attributed to Julius Caesar, who needed to securely send private messages to his confidantes.[9] His solution was to use a simple encrypting scheme, which has come to be known as *Caesar rotation* (or the *Caesar cipher*), that replaces each letter by the one that follows three positions later in alphabetical order. Thus, `A` becomes `D`, `B` becomes `E`, and so forth, up through `W` becoming `Z`. He then did a "wrap-around," with `X` becoming `A`, `Y` becoming `B`, and `Z` becoming `C`. This defines an encrypting function $e\colon \{\mathtt{A}, \mathtt{B}, \dots, \mathtt{Z}\} \to \{\mathtt{A}, \mathtt{B}, \dots, \mathtt{Z}\}$, defined by the table

x	A B C D E F G H I J K L M N O P Q R S T U V W X Y Z
$e(x)$	D E F G H I J K L M N O P Q R S T U V W X Y Z A B C

Clearly, the corresponding decrypting function (which translates ciphertext back into cleartext) is the inverse of the encrypting function. That is,

$$e\big(d(x)\big) = x \qquad \text{and} \qquad d\big(e(x)\big) = x$$

for any $x \in \{\mathtt{A}, \mathtt{B}, \dots, \mathtt{Z}\}$. More succinctly,

$$e \circ d = \mathrm{id}_{\{\mathtt{A},\mathtt{B},\dots,\mathtt{Z}\}} = d \circ e.$$

[8]The very word "cryptography" comes from Greek roots, meaning "hidden writing."

[9]See Suetonius, *Life of Julius Caesar* 56. There is a "folklore" belief that Caesar also used this scheme for transmitting military plans; as plausible as this may be, Suetonius doesn't mention this particular usage.

It's fairly obvious that for Caesar rotation, the corresponding decrypting function $d\colon \{\texttt{A}, \texttt{B}, \ldots, \texttt{Z}\} \to \{\texttt{A}, \texttt{B}, \ldots, \texttt{Z}\}$ is given by

y	A	B	C	D	E	F	G	H	I	J	K	L	M	N	O	P	Q	R	S	T	U	V	W	X	Y	Z
$d(y)$	X	Y	Z	A	B	C	D	E	F	G	H	I	J	K	L	M	N	O	P	Q	R	S	T	U	V	W

For example, suppose that the message to be sent was

```
ATTACK AT NOON.
```

Applying Caesar's encrypting function e to this message, you get the ciphertext

```
DWWDFN DW QRRQ.
```

If an enemy were to see this message and if he didn't know the secret, he'd simply dismiss it as gibberish. Of course, if the message were to be transmitted to Caesar's forces (who had already been told what the encrypting and decrypting methods were), then Caesar's allies would know that the plaintext was originally

```
ATTACK AT NOON.
```

Of course, the security that worked for Julius Caesar wouldn't work so well today. His encrypting/decrypting system is symmetric; once you know the encrypting technique, it's pretty easy to figure out the decrypting technique. Moreover, with modern computers, cracking such a Caesar cipher is a snap. This is another example of why "security through obscurity" doesn't work very well.

5.6.2 Cryptography in Cyber-Commerce

Now let's look at a more modern example—the security of e-business transactions.

Suppose you would like to set up an e-commerce website. As part of the normal way of doing business, you want to accept credit card numbers. However, if a cleartext version of a credit card number is sent from the customer's home computer to the web server that runs the e-commerce site, the odds are very good that the credit card number will be compromised. That's because things never move directly from one point to another on the Internet, but usually via several jumps. If anybody is "listening in" on the client-server conversation at one of these intermediate points, she will then know the customer's credit card number.

So what you need is a way to encrypt a credit card number as ciphertext, such that decrypting the ciphertext is computationally infeasible. In other words, if we let C_{plain} be the set of plaintext credit card numbers and C_{cipher} be the set of ciphertext credit card numbers, we want an *encryption function* $\text{Enc}: C_{\text{plain}} \to C_{\text{cipher}}$ and a decryption function $\text{Dec}: C_{\text{cipher}} \to C_{\text{plain}}$, which is once again the inverse of Enc. This way even if the details of computing Enc were to leak out, it would still be hard for a Bad Guy to compute Dec.

It turns out that such an encrypting/decrypting pair exists. It is based on the fact that it's computationally easy to multiply two huge numbers together, but that nobody knows how to efficiently factor a huge number as a product of prime numbers. In other words, the encrypting method uses the product of two large prime numbers, whereas the decrypting method uses those two large prime numbers themselves.

Warning: If you read that last paragraph too quickly, you might miss an important point. We said that "nobody knows how to efficiently factor a huge number as a product of prime numbers." We didn't say that efficient factoring is impossible. This leaves open the possibility that in the future, somebody might discover an efficient technique for factoring large numbers. This would be a major intellectual achievement, which would probably have the side effect of totally destroying e-commerce (unless somebody comes up with a better scheme for secure submission of personal information such as credit card numbers). □

5.7 More About Functions

We have already seen some uses of functions, e.g., in cryptography. Where else might functions be used in computer and information science? In this section, we give three examples of such usage.

5.7.1 Standard Mathematical Functions

Not surprisingly, there are certain standard functions that appear in computing. The most well-known are the standard mathematical functions; a sampling of which may be found in Table 5.1. You may have encountered some of these functions before in math courses you have previously taken. It should come as no surprise that these functions are found in the standard math library for the UNIX® operating system; they are typeset in a `Courier` font, which is traditional for the typesetting of computer code.

math name	UNIX name	description
$\sqrt{}$	`sqrt`	square root
sin	`sin`	trigonometric sine
cos	`cos`	trigonometric cosine
tan	`tan`	trigonometric tangent
$\sin^{-1}$	`asin`	trigonometric arc (inverse) sine
$\cos^{-1}$	`acos`	trigonometric arc cosine
$\tan^{-1}$	`atan`	trigonometric arc tangent
$\lvert\cdot\rvert$	`fabs`	absolute value

Table 5.1: Some commonly-occuring functions in continuous mathematics

There is another class of functions that arises in computer science, however, which consists of functions not often seen in traditional pre-college math courses. Here is a small sampling:

- The max function. If x and y are numbers, then $\max(x, h)$ is the maximum of x and y. For example, $\max(2.3, -4.2) = 2.3$.
- The min function. If x and y are numbers, then $\min(x, h)$ is the minimum of x and y. For example, $\min(2.3, -4.2) = -4.2$.
- The floor function. If x is a number, then $\lfloor x \rfloor$ is the largest integer that is less than or equal to x. For example, $\lfloor 4.999 \rfloor = 4$.
- The ceiling function. If x is a number, then $\lceil x \rceil$ is the smallest integer that is greater than or equal to x. For example, $\lceil 4.001 \rceil = 5$.

These functions can also be found in the UNIX standard library, but are named `fmax`, `fmin`, `ceil`, and `floor` respectively.

5.7.2 Growth Functions

Functions can be used to describe how quickly things change with time. Perhaps you have heard people discussing *exponential growth*, when discussing population growth. Exponential growth occurs when the amount that a quantity changes at a given time is proportional to how much of the quantity is present at that time. For example, if we're looking at population growth, this says that the number of new babies is proportional to the number of people currently present.

To model exponential growth and to see how it arises, let p_n be the amount present after n time units have elapsed and let a denote the proportionality factor, then

$$p_n = a \cdot p_{n-1} \qquad \text{for } n \geq 1.$$

This is a recursive formula for a sequence. Using the techniques of Section 2.1, we see that the closed form of this sequence is

$$p_n = p_0 a^n \qquad \text{for } n \geq 0,$$

where p_0 is the amount initially present.[10] Note that if $a > 1$, then p_n will grow remarkable quickly as n increases. For example, suppose that we choose $a = 2$ and $p_0 = 1$. Then $p_n = 2^n$, which is the binary sequence that we studied in Example 2.2.

Functions are also used to measure the efficiency of algorithms, a subject that we will discuss in Chapter 8. Suppose that we have an algorithm that solves a particular problem. We often use a *growth function* to measure the efficiency of said algorithm. Such a growth function is typically a function $f: \mathbb{N} \to \mathbb{N}$, with $f(n)$ measuring the cost of using said algorithm to solve this problem when the input size is n. The domain of the growth function is $\mathbb{N}$, since the problem size must be a non-negative number (you can't have a problem size that's negative or not an integer). Moreover, the codomain of a growth function must be a set of non-negative numbers, since cost is never negative; if cost is measured by the total number of operations being performed, then the codomain would typically be $\mathbb{N}$.

We tend to lump some of these growth functions together, further blurring distinctions that may be unimportant at times. The most important boundary line is between polynomial growth functions and functions that grow faster than polynomially (such as functions that grow exponentially). Problems for which we can find an algorithm whose running time is a polynomial function of its input size are said to be *tractable*; problems requiring super-polynomial (e.g., exponential) running time are said to be *intractable*. For example, you'll learn in Chapter 8 that the *bubblesort* algorithm can sort a set of n data items by using about n^2 operations. Since n^2 is a polynomial in n, we see that sorting is tractable.[11] On the other hand, suppose we wanted to list every possible subset of a given set S. This amounts to listing all the elements of $\mathscr{P}(S)$. Since $|\mathscr{P}(S)| = 2^{|S|}$, the

[10] Actually, this involves a slight change from the techniques we learned in Section 2.1, since our sequence starts with $n = 0$, rather than $n = 1$.

[11] We'll also see that you can do even better than bubblesort.

number of subsets grows exponentially in the size of the base set S. Thus this problem is intractable.

To get a feeling for the distinction between polynomial and superpolynomial growth, take a look at Figure 5.8.

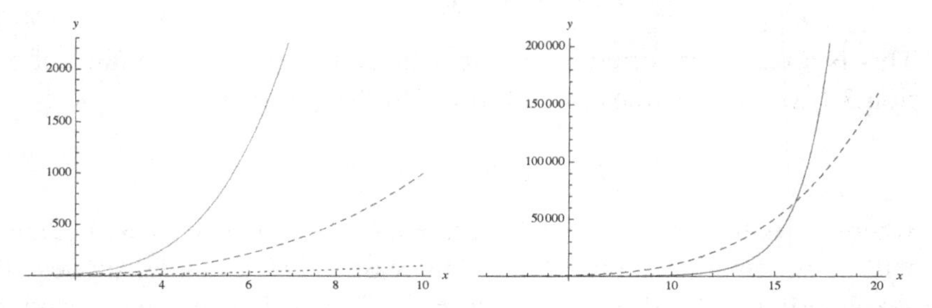

Figure 5.8: Comparison of growth functions

- On the left-hand side, we have graphed the curves $y = x^2$ (dotted line), $y = x^3$ (dashed line), and $y = x^4$ (unbroken line). You can see how the curve $y = x^k$ gets steeper and grows (much) faster as k increases.

- Now look at the right-hand side. The dashed curve is the graph of $y = x^4$, and the unbroken curve is the graph of $y = 2^x$. Although the polynomial curve $y = x^4$ grows faster initially, eventually there's a crossover point, after which the exponential curve $y = 2^x$ grows *much* faster than the polynomial curve.

Remark. Exponentially-increasing functions can be either very bad or very good. We just saw that if the cost of solving a problem increases exponentially with the problem size, then the problem is inherently intractable. On the other had, sometimes things can *improve* exponentially, one example being the exponential improvement in computational ability described by Moore's law, see page 105. □

5.7.3 Functions in Program Construction

Just as business letters, shopping lists, textbooks, newspapers, Tweets®, poems (and so forth) are written in "natural" languages (such as English, Greek, Hebrew, Russian, and Spanish), computer programs are written using *programming languages* (such as C, C++, Java, Lisp, and Perl). Most programming languages allow programmers to create and use functions.

As a result, functions are ubiquitous in the design and implementation of computer programs.

To begin with, functions are the main building block for many computer programming languages. Let's focus our attention on C++, to be specific. Every executable C++ program will have a function named `main`, which is the starting point for program execution. It has the form[12]

```
int main()
{
  ...
}
```

where the ellipsis is replaced by whatever the program is really supposed to do.

But this is only the beginning. As mentioned earlier, a large program is often written as a suite of cooperating functions. To illustrate this point, let's think about a program that repeatedly

- gets input data from some source,
- processes this data, getting some kind of result, and then
- does something with said result (perhaps writes it to the screen or to some output file).

It might look something like this:

```
int main()
{
   do_initialization();
   do {
      put_result(process_data(get_input()));
   } while (more_to_process());
   do_cleanup();
}
```

Note that this `main` function calls additional functions:

- `do_initialization` does whatever initial setup is necessary.
- `get_input_data` gets a data set from some source. This data source might be the keyboard, an existing computer file, the Internet, etc.

[12]If you're wondering about the empty parentheses, that's a C++ syntax detail.

- `process_data` does whatever processing is needed to the input data.
- `put_result` takes the result of the previous step, and puts it someplace. This might be the computer screen, a computer file, the Internet, etc.
- `more_to_process` determines whether or not there is more data to process.
- `do_cleanup()` does whatever final housekeeping is needed before the program can cleanly exit.

There's also the `do/while` construct, which simply says to execute the loop "body" over and over, as long as the `while`-condition is true.

It is worth noting that except for a few small additions (which we're omitting, in the interest of trying to keep things as simple as possible), this `main` function is syntactically correct. Once the six functions mentioned above have been defined, we will then have a C++ program that solves our problem. Moreover, this program is *very* general; this program can solve any problem that can be described by the data flow given above, which covers a wide variety of data-processing applications. Of course, should any of these six functions prove to be too difficult to directly define, they themselves can be defined in terms of simpler functions.

This decomposition process can continue until we are left with easy-to-write functions. Once all the functions are written, we now have a complete system.

Such an approach lends itself well to testing. At any stage in the process, we have a working system; however, the system will be incomplete until all the functions have been fleshed out. (In particular, this might mean that some features could be missing.) So as each function is written, it can be added to the current working system. If the system still works properly after adding the newly-written function, well and good! If not, there's a good chance that the problem lies in the newly-written function, either in its implementation or in how it communicates with the rest of the system.[13] So this gives us a place to look when trying to debug the system.

The heart of this program is the body of the `do/while` block, which consists of one statement, namely

```
put_result(process_data(get_input()));
```

[13]There are other possibilities as well, dealing with the overall design of the system. Such matters are outside the scope of this course.

now contains only one statement. Note that this is function composition. In fact, if we let f = `get_input`, g = `process_data`, and h = `put_result`, then this statement is simply $g \circ h \circ f$. So once again function composition plays an important role.

5.8 An Application: Secure Storage of Passwords

Let's look at the problem of how to securely store user passwords on a computer.[14] In particular, we'll look at a simplified version of how the UNIX operating system handles passwords. The designers of the system created an *encryption function* $f : S \to S$, where S denotes the set of all possible character strings. This function must satisfy the following requirements:

1. It must be an injection.
2. It must be easy to compute.
3. The bijective function $f : S \to \text{Range}(S)$ must be hard to invert.[15] In other words, given an encrypted password e, computing $f^{-1}(e)$ must be prohibitively expensive.

When a user with login ID u chooses a password p, the system computes the encrypted value $e = f(p)$ and stores it, along with the login ID u, in a well-known file that anybody can read.[16] Now suppose that somebody claiming to be the user u wants to log in on the system. She presents the user ID u, along with a purported password $\tilde{p}$, to the login program. This program computes $\tilde{e} = f(\tilde{p})$. Of course $\tilde{e} = e$ happens if and only if $f(\tilde{p}) = f(p)$; since f is an injection, the latter occurs if and only if $\tilde{p} = p$. In short, the system will first compute $f(\tilde{p})$, and then compare this value with u's stored encrypted password e. If they match, then u presented the correct password, and u is admitted into the system; if they do not match, then the incorrect password was given.

We can now look at the roles played by the three conditions on f:

1. If f is not injective, then there exist choices x and $\tilde{x}$ such that $f(x) = f(\tilde{x})$. This means that if the proper password is x and

[14] Some of the ideas contained here are similar to those in Section 5.6, where we discuss cryptography.

[15] We're being a bit sloppy here, in using the same letter f to denote two different functions, the only difference being in their codomains.

[16] This is the file `/etc/passwd`, in case you're really interested.

the user types in the wrong password $\tilde{x}$, the system will not be able to tell the difference between them, and so the user will be admitted with the wrong password.

2. If f cannot be computed quickly, it will take a long time for a legitimate user to log in.

3. If f^{-1} can be computed quickly, then a Bad Guy could look at any encrypted password e for some user u and compute the u's actual password $p = f^{-1}(e)$, totally compromising the system's security.

We close this example with a couple of remarks:

Remark. One example of a prohibitively expensive approach to computing f^{-1} is *exhaustive search.* Given an encrypted password e, we could compute $f^{-1}(e)$ by examining every single $x \in S$ and seeing whether $f(x) = e$; once a match x has been found, then $f^{-1}(e) = x$. However, the cost of doing so is proportional to $|S|$, since we might conceivably need to examine every element of S; although S is finite, its size is huge. To give you an idea how big S can be, let's suppose that a password is required to contain between four and eight characters, and that we allow any printable character to appear in a password. The standard ASCII[17] character set has 95 printable characters.[18] Using the techniques that we will learn in Chapter 6, it is easy to see that the total number of such passwords is $\sum_{j=4}^{8} 95^j = 6{,}704{,}780{,}953{,}650{,}625$, or about 6.7 quadrillion passwords. Suppose that we could test one billion passwords per second. A quick calculation shows that it would take approximately 0.212607 years to try all the possibilities, which is about 77.6 days. Unless the information you're storing is *very* valuable, most Bad Guys wouldn't want to invest that kind of time in cracking your password. □

Remark. The technique that we have described is not foolproof, because it requires the system's users to have a little bit of sophistication.

What happens if you're lazy, and you use only letters and digits in your password? In that case, there are 221,919,451,335,856 possible passwords. The Bad Guy could test these 200 trillion (or so) passwords in about 2.5 days. If you were really lazy, and only used lower case letters, this would be 217,180,128,880 (around 217 billion) passwords, which our Bad Guy could test in around 200 seconds.

[17] This is an acronym for the American Standard Code for Information Interchange.

[18] Namely, the 10 digits, the 26 upper case and 26 lower case letters, the space, and various punctuation symbols.

Even worse: Suppose a naive user chooses a word in some natural language (say, English) as his password. A Bad Guy could mount a *dictionary attack* by grabbing a list of all the words in that language, and then compute $f(x)$ for each English word x; if he finds a match, then he's found that user's password. This means that in principle,[19] a Bad Guy could grab a list of all the encrypted passwords and compare each element of that list with the set of all encrypted "English word" passwords. When he finds a match, that user's account will be compromised.

The moral of the story: choose good passwords. □

5.9 Exercises

5.9.1 Which of the following define functions? If not, why not?

(a) For $x \in \mathbb{R}$, find $y \in \mathbb{R}$ satisfying $y = x^4$.

(b) For $x \in \mathbb{R}$, find $y \in \mathbb{R}$ satisfying $y^4 = x$.

(c) Let $\mathbb{R}^{\geq 0}$ denote the non-negative real numbers. For $x \in \mathbb{R}^{\geq 0}$, find $y \in \mathbb{R}$ satisfying $y^4 = x$.

(d) For $x \in \mathbb{R}^{\geq 0}$, find $y \in \mathbb{R}^{\geq 0}$ satisfying $y^4 = x$.

5.9.2 Figure 5.9 on page 192 consists of three diagrams, each of which (potentially) represents a function. The domain and range are represented by the circles D and R. Each of these candidate functions maps a value from the domain D into a value that may or may not be in the range R. For each of these diagrams:

- Is this a function?
- If this is a function, then is it an injection? a surjection?

5.9.3 Answer the following true/false questions. If the answer is false, explain or give a counterexample.

(a) The inverse of every function is a function.

(b) Every relation is a function.

---

[19] The UNIX system has a mechanism (salting) that makes it harder to mount a dictionary attack. Still, this is no excuse for choosing bad passwords.

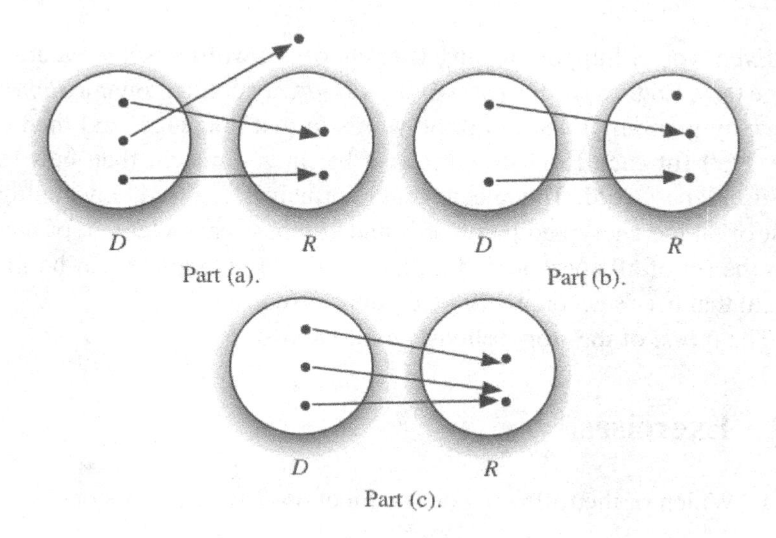

Figure 5.9: Diagrams for Problem 5.9.2

(c) Every function is a relation.

5.9.4 Show that there are two (non-bald) people in the United States of America who have the same number of hairs on their heads.[20] **Hint:** Do a web search to find out how many hairs are on a typical human head.

5.9.5 I own black socks and white socks, which I keep in my sock drawer. The last time I did laundry, I was in a hurry, and I didn't match my socks when I put them away. This morning, I got up before dark and wanted to put on a pair of matching socks. Since I didn't want to wake my spouse, I didn't want to turn on the light. What is the minimum number of socks I should grab from the sock drawer to make sure that I have a matching pair of socks? (This can be answered by common sense or the pigeonhole principle; use the latter.)

5.9.6 Use the pigeonhole principle to determine whether there are two students at your local state university having the same birthday.

[20]The reason we said "non-bald" is to avoid an answer that says, "I know two bald people, and so that solves the problem."

5.9.7 Consider the functions $f, g: \mathbb{R} \to \mathbb{R}$ defined by

$$f(x) = 7x + 3 \text{ for all } x \in \mathbb{R}$$
$$g(x) = 10 \text{ for all } x \in \mathbb{R}$$

Compute the following values:

(a) $(f \circ g)(2)$.

(b) $(g \circ f)(2)$.

(c) $(f \circ f)(2)$.

5.9.8 Consider the functions $f, g, h: \mathbb{R} \to \mathbb{R}$ defined by

$$f(x) = 2x + 3 \text{ for all } x \in \mathbb{R}$$
$$g(x) = x + 1 \text{ for all } x \in \mathbb{R}$$
$$h(x) = x^3 \text{ for all } x \in \mathbb{R}$$

Compute the following values:

(a) $\big(f \circ (h \circ g)\big)(9)$.

(b) $\big((f \circ h) \circ g)\big)(9)$.

5.9.9 (∗)Show that function composition is associative. In other words, let $f: W \to X$, $g: X \to Y$, and $h: Y \to Z$ be functions with the given domains and codomains. Show that

$$h \circ (g \circ f) = (h \circ g) \circ f,$$

5.9.10 Consider the functions $f, g: \mathbb{R} \to \mathbb{R}$ defined by

$$f(x) = 2x + 2 \text{ for all } x \in \mathbb{R}$$
$$g(y) = y^2 - 1 \text{ for all } y \in \mathbb{R}$$

Compute $(f \circ g)(z)$ and $(g \circ f)(r)$.

5.9.11 Find the inverse for each of the following functions.

(a) $s: \mathbb{Z} \to \mathbb{Z}$ defined by $s(n) = n + 1$ for all $n \in \mathbb{Z}$.

(b) $f: \mathbb{R} \to \mathbb{R}$ defined by $f(x) = 2x + 7$ for all $x \in \mathbb{R}$.

5.9.12 This exercise is a reminder that the domain and the codomain are important ingredients in the recipe that determines whether a function is invertible. By analogy with our notations for sets of positive and non-negative sets, let

$$\mathbb{R}^{\geq -1} = \{ x \in \mathbb{R} : x \geq -1 \}.$$

(a) Define $f: \mathbb{R} \to \mathbb{R}^{\geq -1}$ by the rule

$$f(x) = x^2 - 1 \qquad \forall x \in \mathbb{R}.$$

Explain why f is not invertible.

(b) Define $f: \mathbb{R}^{\geq 0} \to \mathbb{R}$ by the rule

$$f(x) = x^2 - 1 \qquad \forall x \in \mathbb{R}^{\geq 0}.$$

Explain why f is not invertible.

(c) Define $f: \mathbb{R}^{\geq 0} \to \mathbb{R}^{\geq -1}$ by the rule

$$f(x) = x^2 - 1 \qquad \forall x \in \mathbb{R}^{\geq 0}.$$

Find f^{-1}.

5.9.13 Suppose that $f: \mathbb{R} \to \mathbb{R}$ is defined by

$$f(x) = 3x - 2 \text{ for all } x \in \mathbb{R}.$$

Let $g: \mathbb{R} \to \mathbb{R}$ be a function such that

$$(f \circ g)(y) = 12y + 7 \text{ for all } y \in \mathbb{R}.$$

Compute $g(y)$.

5.9.14 For each function listed below, answer the following questions:

(a) Give the results of applying three different domain values to the formula.

(b) Is this a function? If not, why not?

(c) Is this injective?

(d) Is this surjective?

(e) Is this bijective?

(f) Does this have an inverse? If so, what is its inverse?

$$f_1 : \mathbb{N} \to \mathbb{Z} \text{ given by } f_1(x) = 2x + 5 \text{ for all } x \in \mathbb{N}$$
$$f_2 : \mathbb{Z} \to \mathbb{N} \text{ given by } f_2(x) = 2x + 5 \text{ for all } x \in \mathbb{Z}$$
$$f_3 : \mathbb{N} \to \mathbb{N} \text{ given by } f_3(x) = 2x + 5 \text{ for all } x \in \mathbb{N}$$
$$f_4 : \mathbb{Z} \to \mathbb{Z} \text{ given by } f(x) = 2x + 5 \text{ for all } x \in \mathbb{Z}$$
$$g : \mathbb{Z} \to \mathbb{Z} \text{ given by } g(x) = 10 \text{ for all } x \in \mathbb{Z}$$
$$h : \mathbb{Z} \to \{\text{all odd positive numbers}\} \text{ given by } h(x) = x^2 + 1 \text{ for all } x \in \mathbb{Z}$$
$$q : \{1, 2, 3, 4, 5\} \to \{11, 12, 13, 14, 15\} \text{ given by } q(x) = x + 10 \text{ for all } x \in \{1, 2, 3, 4, 5\}$$
$$r : \mathbb{Z} \to \mathbb{N} \text{ given by } r(x) = |x| \text{ for all } x \in \mathbb{Z}$$

5.9.15 [SN] Are there any functional relationships in Table A.2 or Table A.3?

5.9.16 Suppose we wish to transmit the secret message

```
THE QUICK BROWN FOX JUMPS OVER THE LAZY DOG.
```

(a) What is the corresponding ciphertext if we use Caesar rotation?

(b) What would our ciphertext look like if we were to use a Caesar rotation by seven letters, rather than three?

5.9.17 Moore's law (see page 105) can be interpreted as saying that computers get twice as good every two years.

(a) How much better will computers get in four years? six years? eight years? in ten years?

(b) (*)How much better will computers get in 100 years? A rough (power of ten) estimate will suffice here, since Moore's law is merely an observation, and not an absolute law of nature.

5.9.18 Floor and ceiling are functions from the real numbers $\mathbb{R}$ into the integers $\mathbb{Z}$, see Section 5.7.1.

(a) Are they injective?

(b) Are they surjective?

(c) Are they bijective?

(d) Are they invertible? If so, give a formula for the inverse function.

Chapter 6

Counting

Don't count your chickens before they are hatched.

Aesop

This chapter introduces the major elements of combinatorics. While combinatorics is about more than just counting, this chapter focuses on counting, which is sometimes called "enumerative combinatorics." The counting that is covered in this chapter generally does not involve counting actual objects, but rather "possibilities." For example, a combinatorics question that you may encounter is: "How many ways can you configure an automobile if there are four possible colors, three different safety options, and a sunroof is optional?" Another question you might encounter is: "If I have 20 immediate Facebook friends and each of those friends has 20 immediate Facebook friends, then how many acquaintances do I have that are reachable by two 'hops' in my social network?" This last question is the subject of Example 6.5.

Why do we care about combinatorics? For one thing, combinatorics provides us with some insight into how the number of possible outcomes grows with respect to the number of decisions, or choices. For example, the number of outcomes that we have per move in a chess game is quite limited, but over several moves the number of possible outcomes grows explosively. Thus we know that, as human-beings, we cannot look at every possible sequence of chess moves in order to determine the best move to make. Coming back to our social network example, while we may not have many immediate friends, if we consider friends of friends to be acquaintances, and so on, then the number of acquaintances that we have will grow very fast as we add new friends. Social networking companies such as LinkedIn exploit this exponential growth, since their social networks allow us to find connections to large numbers of people whom we do not directly know. This exponential growth also largely explains the notion of "six degrees of separation," which says that most people (not just Kevin Bacon[1]) are connected via a chain of friends and acquaintances containing at most six people.

We also care about combinatorics because combinatorics is required to calculate probabilities—and probabilities are often of great practical use in decision-making. The connection between counting and probability will be made clear when we introduce probability in Chapter 7.

6.1 Counting and How to Count

While basic counting is often mastered by four-year old children, there are some aspects that are worth commenting on. First, our counting questions are really about sets. If one asks, "What are all the integers between 5 and 10?", the answer can be expressed by the set $\{5, 6, 7, 8, 9, 10\}$ and similarly, if one asks for all of the male students in a class, the answer can also be represented by a set. Note that in this case, as in many instances, we care not only about the number of objects in the set (i.e., the cardinality of the set), but also about the "names" of the objects in the set. That is, for these counting questions we sometimes want to enumerate the items in the set, not just count them.

The next key issue involves how to count. This topic, although seemingly simple, will be covered in the remainder of the chapter. But in this context we simply mean how one can methodically enumerate a set of items. If we ask you to enumerate the students in your class, you would know how to do this— you would simply create a list of the students. Of course, since we are interested in a set, the ordering of the students in the list would not be important. Nonetheless, most people would order the list either alphabetically or by student seating location, since doing so would ensure than no student is missed or counted multiple times.

In some cases this linear ordering of objects is not the most useful representation. For example, suppose you are asked to select a pair of men's jeans and you know that four styles are available (standard fit, loose fit,

---

[1] The Kevin Bacon game involves trying to link an actor to Kevin Bacon through no more than six connections, where there is a connection between actors if they appear together in a movie.

boot fit, and slim fit) and each style comes in two colors (blue or black). How could you count, or enumerate, the set of possible jean configurations? You could generate a list, which would not be very hard to do with only two characteristics, but this method would not work very well for more complex problems. The best way to enumerate the jean configurations is to use a tree structure, which happens to be a very common structure in computer science (e.g., the directories on your personal computer are organized as a tree). The tree structure, which was introduced earlier in Figure 2.1 on page 67 in the context of mathematical induction, looks like an upside-down tree, with the root at the top and the leaves at the bottom. Each level in the tree corresponds to a different "feature" or "variable" and each branch corresponds to a choice associated with that feature or variable. The tree corresponding to this problem is shown below in Figure 6.1, with the first branch corresponding to the style of the jeans and the second branch corresponding to the color of the jeans. Note that there are a total of eight unique configurations of jeans, each represented by one of the leaves of the tree.

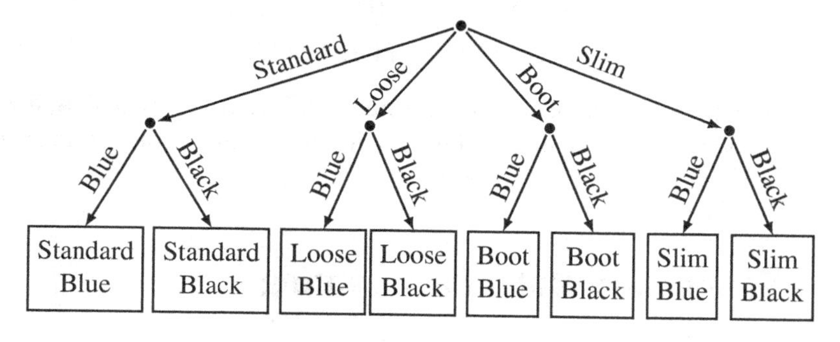

Figure 6.1: Enumeration of jeans configurations using a tree

A tree is one natural way to express the set of possible configurations or outcomes. The same information can be expressed using a table, but only if there are at most two features or choices. Since there are only two choices in this case, a tabular format is appropriate. Table 6.1 expresses the same information as Figure 6.1.

The next section will provide rules to help with the counting process. While these rules are useful and are essential for solving complex problems, simple problems can often be solved by manually enumerating the possible outcomes.

Example 6.1: You are given three coins: a penny, a nickel and a dime. You toss all three high into the air and then record whether they land heads

Color	Jeans Style			
	Standard	Loose	Boot	Slim
Blue	Standard-Blue	Loose-Blue	Boot-Blue	Slim-Blue
Black	Standard-Black	Loose-Black	Boot-Black	Slim-Black

Table 6.1: Enumeration of jeans configurations using a table

up or tails up. Enumerate the possible configurations of heads and tails for these three coins.

Solution: The first thing we need to do is decide how to encode our answer. We choose to encode each individual configuration of the three coins using a triple, where the first position corresponds to the penny, the second corresponds to the nickel, and the third corresponds to the dime. Thus the triple (H,H,T) represents a toss where the penny and nickel land with the head's side facing up and the dime lands with the tail's side facing up. Now we can enumerate the possibilities as the set

$$
\begin{aligned}
C = \{&(H,H,H), (H,H,T), (H,T,H), (H,T,T), \\
&(T,H,H), (T,H,T), (T,T,H), (T,T,T)\}.
\end{aligned}
$$

Note that there are a total of eight configurations. Shortly we will learn how to come up with the cardinality of this set without having to enumerate each item. □

6.2 Elementary Rules for Counting

In this section we introduce two basic rules for counting: the addition rule and the multiplication rule. These rules are primarily introduced through the use of examples. However, we'll introduce some additional terminology before proceeding further. In the jeans configuration example, the choices we had involved the style of the jeans and the color of the jeans. We also referred to these as features or variables, but in the remainder of this chapter we refer to them more generally as just *choices*. Each choice has a number of possible values, or, more formally, a set of *outcomes*. Each choice may ultimately lead to only a single outcome.

6.2.1 The Addition Rule

Before formally describing the addition rule, let's begin with a very simple, but illustrative, example.

Example 6.2: You need to purchase one shirt of any kind. The store has three short-sleeve shirts (red, green, and blue) and four long-sleeve shirts (yellow, purple, pink, and white). How many possible ways are there to choose a shirt?

Solution: Since you can purchase a shirt of any kind, you can purchase either a short-sleeve shirt *or* a long-sleeve shirt. Since there are three ways to pick a short-sleeve shirt and four ways to pick a long-sleeve shirt, there are $3 + 4 = 7$ outcomes (i.e., ways to pick a shirt). □

Most people can solve this problem without any help. However, things can become more complicated when there are many choices and many different kinds of choices. Thus we formalize the reasoning below, by introducing the addition rule, which you may want to think of as the "or" rule.

Addition Rule

If we have two choices C_1 and C_2, with C_1 having possible outcomes O_1 and C_2 having possible outcomes O_2, with $|O_1| = n_1$ and $|O_2| = n_2$, then if the sets O_1 and O_2 are disjoint (i.e., non-overlapping) the total number of outcomes for C_1 *or* C_2 occurring is $n_1 + n_2$. More generally, if we have k choices $C_1, \ldots, C_k$ having $n_1, \ldots, n_k$ possible (non-overlapping) outcomes, then the total number of ways of C_1 occurring or C_2 occurring or ... or C_k occurring is $n_1 + n_2 + \cdots + n_k$.

Since this rule is straightforward, we provide no additional explanation for the standard case, where the sets of outcomes are disjoint. Example 6.2, provided before the addition rule was formally introduced, shows how this rule can be used. Other examples that use this rule are provided throughout the remainder of this chapter. But what if the outcomes are not disjoint? Given our knowledge of sets and the law of inclusion/exclusion from Chapter 1, the answer should be clear—we simply subtract out the elements that overlap, so that they are not counted more than once. Thus, with two choices the number of (unique) outcomes would be $n_1 + n_2 - |O_1 \cap O_2|$. We do not bother to formalize this with a modified addition rule, but rather illustrate the modified addition rule with an example.

Example 6.3: [SN] If you want to choose someone from the social network on page 317 who is a woman (i.e., female) or still lives in their hometown, how many people do you have to choose from? The relevant information is provided in Table A.1 on page 318.

Solution: Let C_1 represent the choice of selecting a woman (Sex $=$ F) from the social network and C_2 represent the choice of a selecting a person who still lives in their hometown. Then according to the data in Table A.1, $n_1 = 15$ and $n_2 = 14$ (i.e., there are 14 entries in Table A.1 for which City $=$ Hometown). In this case the intersection of the outcomes associated with the two choices, $O_1 \cap O_2$, represents the number of women that still live in their hometown and according to Table A.1 this value, $|O_1 \cap O_2|$, equals 9. Thus, if we can choose a person who is either a woman *or* still lives in their hometown, then we can choose from $15 + 14 - 9 = 20$ people. □

6.2.2 The Multiplication Rule

Whereas the addition rule is used when we have several possible choices but will make only one of them, the multiplication rule is used when we have several choices and must make *all* of them. In this case we will need to multiply the number of outcomes for each choice to get the total number of possible outcomes. Figure 6.1, which enumerates the number of configurations of jeans, illustrates this multiplication process. At the first choice point we have four outcomes, corresponding to the four possible styles of jeans. We then must choose the color, and since there are two colors to choose from, the number of possible outcomes then grows by a factor of two. This example suggests and motivates a general rule for counting, the multiplication rule, which is defined below.

Multiplication Rule

If we have two choices C_1 and C_2, with C_1 having possible outcomes O_1 and C_2 having possible outcomes O_2, with $|O_1| = n_1$ and $|O_2| = n_2$, then the total number of possible outcomes associated with C_1 *and* C_2 occurring is $n_1 \times n_2$. More generally, if we have k choices $C_1, \ldots, C_k$ having $n_1, \ldots, n_k$ possible outcomes, then the total number of ways of C_1 occurring and C_2 occurring and ... and C_k occurring is $n_1 \times n_2 \times \cdots \times n_k$.

While the addition rule can be thought of as the "or" rule, the multiplication rule can be thought of as the "and" rule. The multiplication rule for counting is only applicable if *all* the choices must occur. More specifically, if there are k choices, all of $C_1, C_2, \ldots, C_k$ must occur. This rule is applicable to the jeans example described earlier since one must specify the style of jeans *and* the color of the jeans. In English, the "clue" that this rule is applicable has to do with the use of the word "and."

Now let us apply this to the jeans example discussed earlier. In mathematics it is important to properly describe the problem and use the correct terminology, so we will pay special attention to this in the example. Very often if a mathematical problem is formulated properly, the actual solution is trivial.

Example 6.4: You are asked to select a pair of men's jeans and you know that four styles are available (standard fit, loose fit, boot fit, and slim fit) and each style comes in two colors (blue or black). How many possible jean configurations are there?

Solution: We begin by formulating the problem. First we define the choices:

$$C_1 = \text{"choose the jeans style"},$$
$$C_2 = \text{"choose the jeans color"}.$$

Then we list the outcomes:

$$O_1 = \{\text{standard fit, loose fit, boot fit, slim fit}\},$$
$$O_2 = \{\text{black, blue}\}.$$

Now determine the cardinalities of the sets:

$$n_1 = |C_1| = 4,$$
$$n_2 = |C_2| = 2.$$

Now we apply the multiplication rule

$$\text{Total number of outcomes} = n_1 \times n_2 = 4 \times 2 = 8.$$

Note that in this case we were not really asked to enumerate the outcomes, but only to determine the total number of outcomes. This is typical for counting problems. When solving counting problems like this you would typically skip the step where you enumerate the outcomes. We only enumerated the outcomes here for completeness. □

You should be able to see the connection between the multiplication rule for counting and the tree structure in Figure 6.1: the number of leaves in the tree is $n_1 \times n_2$. You should also be able to relate the enumeration process to your knowledge of set theory. In this case the enumeration of the outcomes could be represented by the Cartesian product of the outcomes associated with sets O_1 and O_2. Recall that the cardinality of the Cartesian product $A \times B$ of a set A and a set B, as stated previously in equation (1.1), is:

$$|A \times B| = |A| \cdot |B|.$$

So the fact that the total number of outcomes is given by

$$|O_1 \times O_2| = |O_1| \cdot |O_2|$$

should not surprise us.

The best way to learn about combinatorics is with examples, so we include several additional examples in this section. When trying to solve these problems, remember that coming up with the problem statement is as important as the actual solution.

Example 6.5: [SN] If we make the simplifying assumption that everyone in Facebook has 20 friends, then what is the maximum number of acquaintances reachable in the social network via the friend relation in exactly two "hops"?

Solution: Given the simplifying assumption, everyone has 20 "direct" friends (i.e., friends that can be reached in one hop). Since each of those friends has 20 friends, there are at most $20 \times 20 = 400$ acquaintances reachable in exactly two hops. But in our answer why must we say "at most" 400 acquaintances? Because people usually have friends in common and hence not all of the 400 acquaintances will be distinct.

Example 6.6: You are asked to flip a coin two times and to record the outcome (head or tail) for each flip. How many possible outcomes are there? Then enumerate the possible outcomes.

Solution: There are two choices, C_1 and C_2, corresponding to the two coin flips. C_1 *and* C_2 must occur, so the multiplication rule applies. Each choice has two possible outcomes, thus $n_1 = 2$ and $n_2 = 2$. Thus by the multiplication principle of counting, there are $2 \times 2 = 4$ possible outcomes.

The possible outcomes can be represented using pairs. For instance, the pair (H,T) represents a head on the first flip and a tail on the second flip. The possible outcomes are then {(H,H), (H,T), (T,H), (T,T)}. Again, note that you could generate this by calculating the Cartesian product of the individual sets of outcomes. □

Example 6.7: You are asked to flip a coin five times and to record the outcome (head or tail) for each flip. How many possible outcomes are there?

Solution: This example differs from the previous one only in that there are five choices instead of two. For each choice there are two possible outcomes, so the total number of outcomes is $2 \times 2 \times 2 \times 2 \times 2 = 2^5 = 32$. □

Example 6.8: You are asked to flip a coin n times (where n is a variable) and to record the outcome (head or tail) for each flip. How many possible outcomes are there?

Solution: This example differs from the previous two only in that there is a variable number of coin flips. The solution must therefore express the answer in terms of this variable. Hopefully it will be clear, especially after the previous two examples, that there are 2 choices per coin flip and since there are n flips, the total number of choices will be 2^n. □

Note that if we view the number of outcomes as a function of the number of coin flips in this example, then the function exhibits exponential growth. Exponential functions grow very rapidly, as we saw in Section 5.7.2.

In our earlier discussion on sets in Chapter 1, we mentioned that if a set has n elements then its power set has 2^n elements. We motivated this with a few examples, but did not have the background to provide a good justification for this formula. We now have the necessary background. In fact, the previous example provides the justification, once we understand how to view the number of items in the power set as a counting problem. Given a set A with n elements, the power set is defined to be the set of all possible sets generated from these elements. This can be viewed as a counting problem where the choices we have involve whether or not each element is included or excluded when generating a set from the n original elements. So, we can choose to include or exclude the first element, we can chose to include or exclude the second element, ..., and we can choose to include or exclude the nth element. If we enumerate all possible choices, we will generate all items in the power set. Thus, since we have two choices for each element (include or exclude) and there are n elements, there are 2^n choices, as we showed in the previous example.

An example will show exactly how the power set is equivalent to a counting problem. First, imagine that we have a set A, where $A = \{a, b\}$, and we want to know the number of items in A's power set. We have two possible choices for element a (include or exclude) and also for element b (include or exclude). Based on counting and the multiplication rule, the $2 \times 2 = 4$ choices that we can make for elements a and b are (exclude, exclude), (include, exclude), (exclude, include), and (include, include), which correspond to the four elements $\emptyset$, $\{a\}$, $\{b\}$, $\{a, b\}$ of A's power set.

Example 6.9: You play a lottery where you choose five numbers and each number must be between 1 and 20, inclusive. If a number may be selected more than once, then how many ways can you fill out the lottery card?

Solution: In this example there are five choices, corresponding to the five numbers that you must choose. Each of the five choices must occur, so the multiplication rule applies. Each choice has twenty possible outcomes (i.e., you pick a number between 1 and 20). Therefore, there are $20 \times 20 \times 20 \times 20 \times 20 = 20^5 = 3{,}200{,}000$ possible ways to fill out the lottery card. Note that in this example we assume that lottery numbers can be used more than once. □

Example 6.10: You play a lottery where you choose five numbers and each number must be between 1 and 20, inclusive. The numbers are chosen by the lottery commission from a bin and once a number is chosen it is discarded and cannot be chosen again. In how many ways can you fill out the lottery card?

Solution: This example is identical to the previous one except that a number cannot be chosen more than once. This means that the number of possible outcomes for each choice is progressively reduced by one. Assuming the five choices $C_1 \ldots C_5$ are numbered such that C_1 corresponds to the first number selected and C_5 to the last number selected, then the number of outcomes for C_1 is 20, for C_2 is 19, for C_3 is 18, for C_4 is 17 and for C_5 is 16. Thus the number of possible outcomes is $20 \times 19 \times 18 \times 17 \times 16 = 1{,}860{,}480$. Note that this answer is a little more than half of the answer for the previous example. □

The key point from the last example is that the number of outcomes may change over time and may be affected by previous choices. This will be a very common occurrence since, for many problems, once an object is selected, it may not be re-selected.

Example 6.11: How many ways are there to choose one additional class for your schedule if there are five evening classes *and* four day classes?

Solution: In this example the task is to choose one class. You can choose a day class *or* an evening class, so the addition rule applies and there are $5 + 4 = 9$ ways to choose this class. Note that you could be misled by the use of "and" in the question (which was italicized to further add to the confusion) or by the fact that the example occurs in the section on the multiplication rule! The point of this example is that you must pay attention to what is really being asked and not only look at the terms that occur in the sentence. The key part is how the choices are *combined* and in this case it is understood by the context of the question that you can pick a day *or* an evening class. The "and" in this case does not refer to relationship between the choices. □

6.2.3 Using the Elementary Rules Together

In this section, we present additional counting problems that may be solved using a combination of the addition and multiplication rules. However, in some cases the problem can be solved using a single rule by viewing it in a certain way. This will be made clear by some of the examples.

Example 6.12: How many odd three digit numbers are there, assuming that leading zeros are permitted (e.g., the number 007 is valid)?

Solution: There are two nearly identical ways to solve this problem. Both methods involve three choices, related to choosing the first, second, and third digits, where the first digit is the most significant digit (i.e., in the hundreds column). Both approaches make use of the fact that a number is odd if it ends with a 1, 3, 5, 7, or 9.

The most natural method only makes use of the multiplication rule. We simply choose values for C_1 and C_2 and C_3. How many choices are there? For C_1 and C_2 there are no restrictions so there are 10 choices for each, while there are only five choices for the C_3, which represents the "ones" digit. This yields $10 \times 10 \times 5 = 500$ outcomes.

However, we can view this problem in a slightly different manner. We say the number of outcomes is the number of outcomes where the number ends in a 1 *or* 3 *or* 5 *or* 7 *or* 9. That is, we break down the problem into five cases and then determine the number of outcomes for each of these five cases and then add them. In this case there are $10 \times 10 = 100$ outcomes from each case, and so we get $5 \times 100 = 500$ outcomes overall. □

The next counting example is only a bit more complicated than Example 6.12, but assumes that you know the basics about the makeup of a deck of cards and about poker. Since card and poker examples abound in combinatorics and probability, we provide a basic lesson on the subject here.

Facts About Playing Cards and Poker

A deck of cards contains 52 cards. Each card belongs to one of four suits: ♣ (Clubs), ♢ (Diamonds), ♡ (Hearts) and ♠ (Spades) and is associated with one of thirteen denominations: 2, 3, 4, 5, 6, 7, 8, 9, 10, J(ack), Q(ueen), K(ing), A(ce). The clubs and spades are black and the diamonds and hearts are red. Unless otherwise specified, assume that for any example you begin with a complete deck and that as cards are dealt they are not immediately replaced back into the deck. We abbreviate a card using the denomination and then suit, such that 2♡ (or 2H) represents the 2 of Hearts.

In standard (draw) poker you receive 5 cards, called a *hand*. While you can later discard cards and then replace them, for most of our examples we will only consider the initial configuration.

Poker hands have the following names:

- A *pair* in poker (*two of a kind*) means that you have two cards that are the same denomination, such as a pair of fours.
- *Three of a kind* and *four of a kind* are defined similarly.
- A *full house* is a hand containing three cards of one denomination and two cards of another denomination (i.e., three of one kind and a pair of another kind).
- A *straight* means that the cards are in sequential order, possibly in more than one suit.
- A *flush* means that all five of the cards are of the same suit, not necessarily in sequential order.
- Finally, a *straight flush* is a hand in which all the cards are of the same suit and in sequential order, i.e., it's both a straight and a flush.

These hands are ranked as follows (from least valuable to most valuable): one pair, two pairs, three of a kind, straight, flush, full house, four of a kind, straight flush. (Of course, a hand that has none of these is the worst kind of hand; the very worst poker hand is 2, 3, 4, 5, and 7, in more than one suit.)

All four suits are equally important and in poker the suit value is never used to break ties. However, the face values of individual cards

Facts About Playing Cards and Poker (cont.)

are ranked

$$2 < 3 < 4 < 5 < 6 < 7 < 8 < 9 < 10 < \text{J} < \text{Q} < \text{K} < \text{A}.$$

This ordering allows us to break (what might otherwise be) ties. For example, suppose that two players each are holding a full house, with the first player holding 6♡, 6♢, 6♣, 3♠, 3♢ and the second player holding 5♠, 5♢, 5♡, 4♢, 4♣. Since 6 outranks 5, the first player beats the second player.

In poker one only classifies a hand using the highest rank possible. For example, the hand 6♡, 6♢, 6♣, 6♠, 3♢ would always be referred to as *four of a kind* and not *two pair* and, similarly, a hand that qualifies as a *straight flush* would not be referred to as a *flush*.

Example 6.13: How many ways can you draw a flush in poker, assuming that the order of the five cards drawn matters? (We will learn how to relax this assumption in the next section.)

Solution: Since a flush means that all five cards are of the same suit and there are four suits, there are four basic ways to get a flush: all clubs or all diamonds or all hearts or all spades. We can view each of these as outcomes that satisfy the condition of drawing a flush; then our goal is to determine the total number of outcomes of these four disjoint (i.e., non-overlapping) outcomes. The number of ways we can get five of the same suit is the same for all suits, so we only need to compute the number for any suit and then multiply the answer by the number of suits, which is four. Below we arbitrarily select clubs and figure out the number of ways to draw 5 clubs. Note that we start with 13 clubs in the deck. Using the multiplication rule to select 5 cards, below we compute the number of ways to draw five clubs:

$$\text{Number of ways to draw five clubs} = 13 \times 12 \times 11 \times 10 \times 9 = 154{,}440.$$

Therefore, by the addition rule, there are $4 \times 154{,}440 = 617{,}760$ ways to get a flush (if the order that cards are drawn matters). □

6.3 Permutations and Combinations

When counting, we sometimes care about the order of the choices that are made, while other times we do not care about the order. But whether the order matters or not makes a big difference. To see this, imagine that after selecting five cards, you have a hand consisting of the cards 2♡, 3♠, 4♢, 5♣, 6♡. If we do not care about order, then we count this configuration once, regardless of what order the cards were drawn in. However, if we care about the sequence in which the cards were drawn, then there are many sequences that can be formed containing only these cards and all of these sequences should be considered distinct. How many ways can we draw five cards and wind up with the five cards just listed? Using the multiplication rule for counting, there are 5 possibilities for the first draw, 4 for the second, 3 for the third, 2 for the fourth and 1 for the last. This yields $5 \times 4 \times 3 \times 2 \times 1 = 120$ ways. Thus we see that if order does matter, we will always get a bigger number than if order does not matter. In this particular example, the answer we get is 1 if order does not matter and 120 if order does matter. For every problem in combinatorics that you try, one of the first things you should ask yourself is whether order does or does not matter. Sometimes the problem will mention this explicitly, whereas other times you must determine whether order matters or not based on the context of the problem. In all of the examples so far, we have assumed that order *does* matter. If order does not matter, then the answer we come up with must be reduced. We will shortly see how to do this. In combinatorics, order is important for permutations but not for combinations.

6.3.1 Permutations

A permutation of objects is an arrangement where order, or position, does matter. For example, the telephone number `123-456-7890` is not the same as the telephone number `789-012-3456` so it makes sense to talk about permutations when discussing phone numbers. However, for permutations not only does order matter, but each "object" can only be used once in the permutation. As an example, imagine that we have five children and five chairs in a row. Instead of asking how many ways we can place the children into the five chairs, we could simply ask how many permutations are there of the five children. Note that this problem meets the key criteria for a permutation. First, order does matter (in this case it can be inferred from the problem). Second, a child cannot sit in two chairs at once. Using our knowledge of counting, we know by the multiplication rule that there are $5 \times 4 \times 3 \times 2 \times 1 = 120$ permutations. Now what if we have 10 children and want to place them into the 5 chairs? In this case the answer would be $10 \times 9 \times 8 \times 7 \times 6 = 30{,}240$. There is a general formula to calculate permutations, although we could always compute the number from scratch using the multiplication rule.

The number of permutations for r items chosen from a total of n items is denoted $P(n,r)$. The number of permutations for placing 10 children into 5 chairs is $P(10,5)$. We could also say this slightly differently: the number of ways to permute 10 children into 5 chairs is $P(10,5)$. The number of ways to permute 5 children into 5 chairs is denoted $P(5,5)$. The formula for calculating $P(n,r)$ is given by

$$P(n,r) = \frac{n!}{(n-r)!}\,. \tag{6.1}$$

Here, $n!$ (pronounced "n factorial") is defined as the product of the positive integers n down to 1, i.e.,

$$n! = n \times (n-1) \times (n-2) \cdots \times 2 \times 1\,.$$

Although 0 is not a positive integer, it is useful to define[2]

$$0! = 1\,.$$

Here are several examples:

- $0! = 1$.
- $1! = 1$.
- $5! = 5 \times 4 \times 3 \times 2 \times 1 = 120$.
- $10! = 10 \times 9 \times 8 \times 7 \times 6 \times 5 \times 4 \times 3 \times 2 \times 1 = 3{,}268{,}800$.

Using the formula for a permutation, we see that the number of permutations of 10 children in 5 seats is

$$P(10,5) = \frac{10!}{(10-5)!} = \frac{10!}{5!}\,.$$

Rather than computing 10! and 5! and then dividing the two, we can save some effort by crossing out the common elements, getting

$$\begin{aligned} P(10,5) &= \frac{10!}{(10-5)!} = \frac{10!}{5!} = \frac{10 \times 9 \times 8 \times 7 \times 6 \times \cancel{5} \times \cancel{4} \times \cancel{3} \times \cancel{2} \times \cancel{1}}{\cancel{5} \times \cancel{4} \times \cancel{3} \times \cancel{2} \times \cancel{1}} \\ &= 10 \times 9 \times 8 \times 7 \times 6 = 30{,}240. \end{aligned}$$

[2]Many students find the definition "$0! = 1$" to be counter-intuitive, thinking that 0! should be 0. There are several reasons why letting $0! = 1$ is not only permissible, but is also the right thing to do. The simplest reason is that it makes all our formulas work out right when a zero is present.

Note that this is the same answer we computed using the multiplication rule. It is important that you be able to compute these answers from "first principles," but once you have seen enough examples it is acceptable to just use the formula for permutations.

Now that we have performed the calculations and have seen that the permutation formula achieves the same result as solving the problem from first principles, it is not difficult to see how the permutation formula works. We have n objects and need to arrange r of them. There are n ways to pick the first object, $n - 1$ ways to pick the second object, $n - 2$ ways to pick the third object, ..., and $n - (r + 1)$ ways to pick the rth object. This is equivalent to $n!/(n-r)!$ since the $(n-r)!$ removes (i.e., crosses out) from the $n!$ in the numerator all of the values after the $n - (r + 1)$.

You should also note the following, which may help you avoid some computational mistakes. The number of permutations counts the number of ways we can accomplish some task. In the prior example it's the number of ways we can seat the children. Of course, you can't seat the children in $2\frac{1}{2}$ or -3 ways! This illustrates an important "sanity check," namely, that the number of permutations must *always* be a non-negative integer. *In other words, if you get a negative or fractional answer for $P(n, r)$, you need to go back and double-check your arithmetic!*

Example 6.14: If a baseball team has 25 players, how many batting orders are there?

Solution: First, for those not familiar with baseball, a batting order consists of nine players and order does matter (i.e., it matters who bats first, second, etc.) Since a batter cannot bat twice in a lineup, the question asks how many permutations there are for 25 players into 9 spots (the rest of the players are on "the bench"). The answer is:

$$P(25, 9) = \frac{25!}{16!} = 25 \times 24 \times \cdots \times 17 = 741{,}354{,}768{,}000. \qquad \square$$

6.3.2 Combinations

Permutations assume that order matters. For a baseball lineup order does matter, but order does not matter if instead one is selecting members to be on a jury or on a committee—assuming that all members have identical roles. When order does not matter, we care about combinations, not permutations. If we need to choose r items from n and order does not matter, we denote this as $C(n, r)$. Thus, if we have a jury pool of 100 people and need to choose 12 people, the number of combinations would be

$C(100, 12)$, read as "100 choose 12." The formula for computing combinations is given by:

$$C(n, r) = \frac{P(n, r)}{r!} = \frac{n!}{r!(n - r)!} \tag{6.2}$$

Note that, as with permutations, an item cannot be selected more than once. For example, when selecting the jury from a pool of 100 people, we would not want to choose one person more than once.

The only difference between the formula for permutations and combinations is that the formula for combinations has an additional $r!$ in the denominator. Given that we previously concluded that the number of outcomes should be reduced if order does not matter, this makes sense. To see why we must reduce this number by a factor of $r!$, we will continue the baseball example.

If we want to compute the number of ways to choose 9 people from 25, where order does not matter, we get $C(25, 9) = 25!/(9! \times 16!)$. Thus, the number is reduced by a factor of $9!$ when compared to the number of permutations. The reason is quite simple: $9!$ of the individual outcomes when order does matter become equivalent when order does not matter. To see this, imagine you have already selected 9 people. How many ways can you order them? You can order them in $9 \times 8 \times 7 \times 6 \times 5 \times 4 \times 3 \times 2 \times 1 = 9!$ ways. In general, r items can be ordered in $r!$ ways, which is why you must divide by $r!$ when order does not matter.

Example 6.15: How many ways can a teacher choose 4 students out of the 20 in her class?

Solution: Since there is no indication that order matters, we assume that order does not matter. Therefore the answer is:

$$C(20, 4) = \frac{20!}{4! \times 16!} = \frac{20 \times 19 \times 18 \times 17}{4!} = 4{,}845. \qquad \square$$

Of course, the number of ways we can choose something must always be a non-negative number. So the advice we gave you in the previous section (for computing permutations) also holds when you are computing combinations: *if you get a negative or fractional answer for $P(n, r)$, you need to go back and double-check your arithmetic!*

Example 6.16: How many ways can a university choose a freshman class president and vice-president if there are 250 freshman students and all are eligible?

Solution: The first question we need to ask here is whether order matters, which will tell us whether we want to compute combinations or permutations. The answer here is that effectively order does matter, since there are two distinct roles to be filled. That is, there is a big difference between the case that Ben Smith is president and Mary Johnson is vice-president and the case that Mary Johnson is president and Ben Smith is vice-president. Since these two cases are different, we must effectively consider the order. Since there are 250 choices for president and then 249 choices for vice-president, the answer is $250 \times 249 = 62{,}250$, which is the same as $P(250, 2)$. □

Example 6.17: A typical telephone number has 10 digits (e.g., `555-817-4495`), where the first three digits are known as the *area code* and the next three as the *exchange*. Answer the following questions:

(a) Assuming *no* restrictions, how many possible (three-digit) area codes are there?

(b) When area codes first appeared, their middle digit had to be either a `0` or a `1` (this is no longer a requirement). How many possible (three-digit) area codes were there when this restriction was in place?

(c) Assuming no restrictions whatsoever, how many possible values are there for the full 10-digit phone number?

(d) If the only restriction is that no digit may be used more than once, how many possible 10-digit phone numbers are there?

Solution:

(a) With no restrictions there are $10 \times 10 \times 10 = 1{,}000$ area codes.

(b) If the middle digit must be `0` or `1` then there are $10 \times 2 \times 10 = 200$ area codes.

(c) With no restrictions there are $10^{10} = 10{,}000{,}000{,}000$ phone numbers.

(d) If each digit can only be used once there are $10 \times 9 \times \cdots \times 1 = 10! = 3{,}628{,}800$ phone numbers. □

Example 6.18: A poker player is dealt a hand of 5 cards from a freshly mixed deck. Answer the following questions concerning the number of ways each of the following hands may be dealt. For each, do not worry about possibly also getting a flush at the same time.

(a) How many ways can one be dealt two pairs?

(b) How many ways can one be dealt three of a kind?

(c) How many ways can one be dealt a full house?

Solution:

(a) To calculate the number of ways that two pairs can be dealt, we need to identify the two denominations that correspond to the two pairs. There are $C(13, 2)$ ways of doing this. Then for each of the two denominations we must choose 2 of 4 specific suits, which can be done in $C(4, 2)$ ways. Finally we need to select the card that is not involved in the two pairs. There are 11 denominations left to choose from (i.e., the 13 denominations minus the 2 used in the pairs) and for each of these 11 denominations there are 4 suits. So, the final answer is that there are

$$C(13, 2) \times C(4, 2) \times C(4, 2) \times 11 \times 4 = 123{,}552 \text{ ways.}$$

(b) To get three of a kind, we first can choose the denomination. There are 13 ways to choose this. Once we have the denomination, we must choose 3 of the 4 available suits, which can be done $C(4, 3)$ ways. Then we must choose the remaining two cards. These cards cannot be of the same denomination used for the three of a kind, so there are 12 denominations to choose from. The only other restriction is that the two cards cannot be of the same denomination since then we would have a full house. There are $C(12, 2)$ ways to choose 2 different denominations. Then for each of these 2 denominations there are 4 choices, corresponding to the 4 suits. Thus the final answer is that there are

$$13 \times C(4, 3) \times C(12, 2) \times 4 \times 4 = 54{,}912 \text{ ways.}$$

(c) A full house requires three of a kind and also two of a kind. We can choose the denomination with three of a kind in 13 ways and then we can choose the three specific cards in $C(4, 3)$ ways. Then we can choose the denomination with the two of a kind in 12 ways and choose the two specific cards in $C(4, 2)$ ways. Thus the final answer is:

$$13 \times C(4, 3) \times 12 \times C(4, 2) = 3{,}744 \text{ ways.} \qquad \square$$

Sometimes the key to solving a counting problem is in properly decomposing the problems into subproblems, solving the subproblems, and then combining the solutions to get the final result. The next example makes this clear. It appears quite simple but is often solved incorrectly.

Example 6.19: How many different ice cream cones can you make if there are two types of cones available (waffle and sugar), three flavors (chocolate, vanilla, and strawberry), and you can have either a one-scoop or two-scoop cone?

Solution: First, note that a one scoop cone is very different from a two scoop cone, since a two scoop cone involves more flavor choices due to the additional scoop. Thus, it is *incorrect* to say that the answer is $2 \times 3 \times 2 = 12$ because there are two types of cones and three flavors, with two choices for the number of scoops. The proper way to solve this problem is to find the number of one-scoop cones and two-scoop cones and then use the addition rule to add these numbers together:

- There are $2 \times 3 = 6$ one-scoop cones.
- There are $2 \times 3 \times 3 = 18$ two-scoop cones.

So the total number of one-scoop and two-scoop cones is $6 + 18 = 24$. □

Example 6.20: How many distinguishable ways are there to arrange the letters in the word `MISSISSIPPI`?

Solution: This question is different from the other permutation and combination questions we have seen thus far. Based on the multiplication rule, we can arrange the letters in $11!$ ways. However, not all of these ways are distinguishable since some letters repeat (`S` and `I` each occur 4 times and `P` occurs 2 times); when a letter repeats, we are over-counting the number of indistinguishable configurations if we treat these letters as if they were all different. For example, let's say we place all of the 4 instances of `S` at the end of the word. With the naïve method we are counting this one distinguishable configuration $4 \times 3 \times 2 \times 1 = 24$ times instead of only once, since there are 4! ways to permute these 4 occurrences of `S`. There are two approaches that we can take to answer this question. Either we can calculate things naively and then correct for the fact that some of the permutations are indistinguishable, or we can calculate things to avoid the problem in the first place. For the sake of illustration we employ both approaches, but you should feel free to use whichever approach that you prefer.

- *Approach 1*: The naïve answer is 11!. We then account for the fact that we have 4 instances of `S` by dividing this answer by 4!. Since we have 4 occurrences of `I` and 2 of `P`, we would also divide by 4! and 2!. Thus, our final answer is

$$\frac{11!}{4! \times 4! \times 2!} = 34{,}650.$$

- *Approach 2*: In this approach we view this as a fill-in-the-blank problem with 11 blanks and then carefully see how many ways we can fill in the blanks. We view this as a combination problem, but here we choose the blanks to fill in, not the letters to use. The best way to follow this is to work through the example. We start with the 11 blanks:

  ___ ___ ___ ___ ___ ___ ___ ___ ___ ___ ___

  We then start assigning letters to the blanks. We will pick the easiest letters first (i.e., the ones with the fewest occurrences). How many ways can we assign the 1 `M` to the blanks above? The answer is $C(11, 1) = 11!/(10! \times 1!) = 11$. As we expect, there are 11 ways. Once this is done, we next decide how many distinct ways to assign the 2 `P`'s. Since there are only 10 slots left, the answer is $C(10, 2)$. Note that because the 2 `P`'s are indistinguishable, we use $C(10, 2)$ instead of $P(10, 2)$. We then assign the 4 `S`'s to the 8 remaining spots in $C(8, 4)$ ways. Then we assign the 4 `I`'s to the 4 remaining spots in $C(4, 4)$ ways (note there is only one way to do this since there are 4 `I`'s and only 4 spots left). The total number of ways is then

$$\begin{aligned} C(11, 1) \times C(10, 2) \times C(8, 4) \times C(4, 4) &= 11 \times 45 \times 70 \times 1 \\ &= 34{,}650. \end{aligned}$$ □

Example 6.21: A binary number is comprised of only `0`'s and `1`'s. How may 7 digit binary numbers have exactly 3 `0`'s and 4 `1`'s?

Solution: This problem is nearly identical to the problem in the previous example, except that here we have digits rather than letters. This problem can be solved using either of the two approaches presented in the previous example. We solve this problem using approach 2. Since there are 7 spots, we can place the `0` into 3 of these slots in $C(7, 3)$ ways. There are then $C(4, 4)$ ways to place the 4 `1`'s. Thus there are:

$$C(7, 3) \times C(4, 4) = 35 \times 1 = 35 \text{ binary numbers.}$$ □

6.4 Exercises

6.4.1 Your school's Computer Club has 25 members, of which 5 are freshmen, 5 are sophomores, 10 are juniors and 5 are seniors.

(a) How many ways can a club president be selected if freshmen are ineligible to be president?

(b) How many ways are there to select two seniors to serve on your school's Activity Council?

(c) How many ways are there to choose a club president and vice-president if the president and vice-president cannot be from the same class and freshmen cannot hold either position?

(d) How many ways are there to choose a club subcommittee with 25 members?

6.4.2 There are 10 people at a party. If every person shakes hands with everyone else at the party, then how many total handshakes take place?

6.4.3 A new car can be ordered with a choice of 10 exterior colors, 7 interior colors, with or without air-conditioning, with or without power steering, and with or without a power package. Assuming all possible combinations are allowable, how many possible versions of this car can one order?

6.4.4 You toss two standard six-sided dice, one red and one blue. How many outcomes are there where the red die has a higher value then the blue die?

6.4.5 You toss two standard six-sided dice. How many ways can you toss a seven (i.e., the sum of both rolls is a 7)? How many ways can you toss a two (i.e., a "snake eyes")? Note that by convention when we work on dice problems we always count rolls such as (4,3) separate from (3,4).

6.4.6 Compute the value of the following expressions and simplify them as much as possible:

(a) $P(5, 2)$

(b) $C(5, 2)$

(c) $P(100, 1)$

(d) $C(100, 1)$

(e) $P(n, n)$

(f) $C(n, n)$

(g) $C(n, 0)$

(h) $C(n, 1)$

6.4.7 A saleswoman must visit all of the ten cities that fall within her sales area. Assuming that there is a road connecting all pairs of cities and the saleswoman starts in New York City (one of her ten cities), how many possible routes are there that have her visit all ten cities without visiting any city twice?

6.4.8 The Mets and the Cardinals play a 5-game series in the first round of the Major League playoffs. The first team to win 3 games wins the series and at that point the series terminates. If we list outcomes from the Mets' perspective, then the series outcome WWLW would indicate a series win in 4 games, with a loss in only the third game.

(a) Enumerate all possible series outcomes from the Mets' perspective. How many such outcomes are there?

(b) Enumerate the series outcomes that have the Mets winning the series. How many such outcomes are there?

(c) What fraction of the total number of possible outcomes have the Mets winning the series?

(d) List the series outcomes that correspond to the Mets losing the first game and then winning the series?

(e) If the Mets lose the first game, then what fraction of the total outcomes have them winning the series?

6.4.9 A class has 15 women and 10 men.

(a) How many ways are there to choose one class member to take attendance?

(b) How many ways are there to choose two people to clean the board?

(c) How many ways are there to choose one person to take attendance and one person to clean the board?

(d) How many ways are there to choose one person to take attendance and one person to clean the board if both jobs cannot be filled with people of the same sex?

(e) How many ways are there to choose a person to erase the board and a person to take attendance if both of the jobs must be filled with people of the same sex?

6.4.10 In the United States social security numbers are issued to track individuals for taxation purposes. These numbers are comprised of 9 decimal digits. Calculate the number of possible social security numbers for each part below, assuming only the stated restrictions.

(a) No restrictions.

(b) The first digit is `0`.

(c) The first and last digits must match.

(d) When interpreted as an integer the value is less than a million.

(e) When interpreted as an integer the value is greater than or equal to a million.

(f) No digit is repeated.

(g) No digit is repeated and the value of successive digits always increases.

6.4.11 A poker player is dealt a hand of five cards from a freshly mixed deck (the order that the cards are dealt does not lead to different hands).

(a) How many different hands will have four aces?

(b) How many different hands will have four of a kind?

(c) How many different hands will have a royal flush (a royal flush is an Ace-high straight flush, i.e., it contains an Ace, King, Queen, Jack and 10, all of the same suit).

6.4.12 Your friend has 15 magazines and says you can borrow any 5 magazines. How many different possible collections of magazines could you choose?

6.4.13 You are going on a vacation to Bermuda and ask your friend if you can borrow some books to read on the beach. Your friends says "Sure, I have five books, take as many as you like." Assuming that you take at least one book, how many different choices do you have?

6.4.14 Forty slips of paper numbered `1` through `40` are placed in a hat and two are drawn out. How many different (unordered) pairs of numbers can be drawn?

6.4.15 If the automobile license plate must be of the form `XXX-DDDD` where `X` is a capital letter and `D` is a digit (0–9), then how many unique license plates are there?

6.4.16 Suppose there are ten applicants, seven men and three women, for a job. The men and women will be interviewed in a random order. How many different ways can they be interviewed?

6.4.17 How many distinguishable arrangements of the word `RADAR` are there?

6.4.18 How many ten-digit binary numbers have exactly three `1`'s?

6.4.19 There are ten students in a class. The instructor is quite dictatorial and insists on telling everyone where to sit. If there are 15 chairs in the classroom, then how many possible seating assignments can the instructor come up with?

6.4.20 A store stocks 20 different shirts and 12 different ties. You want to purchase three different shirts and two different ties. How many ways can you make this purchase? You need not simplify your answer.

6.4.21 **[SN]** Grace, who is one of the most popular people in our social network (page 317) with seven friends, decides to host a party. She sends an invitation to all of her friends and each can accept or decline the invitation. Since Grace does not want to be perceived as being unpopular, she decides to cancel the party if less than four of her friends accept the invitation. How many possible guest lists are there assuming that the order of people on the guest list does not matter? *Show your work but also simplify your answer to a final numerical result.*

6.4.22 [SN] Suppose in the prior exercise Grace was not insecure and would have the party no matter how many people declined her invitation. In this case how many possible guest lists would there be?

6.4.23 Your very limited music collection consists of 3 Rock CDs, 2 Classical CDs, 3 Pop CDs, and 1 Country CD. If you arrange your CDs such that all CDs belonging to the same genre are together, then how many possible arrangements are there?

6.4.24 ($*$) How many ways are there to seat ten people at a round table if we don't care about the absolute position of anyone, only their position relative to the others (i.e., having everyone shift one position to the right does not yield a different configuration)?

6.4.25 ($**$) A bag has 32 balls—8 orange, 8 white, 8 red and 8 yellow. All balls of the same color are indistinguishable. A juggler randomly picks three balls from the bag to juggle. How many possible groupings of balls are there?

Chapter 7

Probability

> It is a truth very certain that when it is not in our power to determine what is true we ought to follow what is most probable.
>
> René Descartes,
> *Discourse on Method*

In this chapter we will learn to calculate the likelihood, or probability, that a specific "event" occurs. For example, if the action is flipping a coin then the event could be "tossing a head." Most of us already have a good intuitive notion of probability and therefore know that, assuming a fair coin, the probability of "getting a head" is $\frac{1}{2}$ or 0.5. In this chapter we will formalize this notion of probability so that much more complex probability questions may be answered. For example, by the end of this chapter you will be able to determine the probability that four flips of a coin will lead to all heads or, alternatively, exactly two heads and two tails. Potentially of more interest to some of you, we will also learn to calculate the probability of drawing a flush in poker or of winning a lottery such as Pick Six, PowerBall or MegaMillions.

Knowing the basics of probability theory is extremely useful and, perhaps more than any topic covered in this book, probability has obvious applications to everyday life. The most obvious application of probability theory is to gambling and other games of chance—and these are heavily represented in the examples in this chapter. However, at least in an informal sense, we routinely use probability in our everyday lives. For example, we will bring an umbrella to school only if the probability of rain (i.e., our estimation of it) is above some threshold value. In this chapter we will learn to *combine* probabilistic information in a mathematically justifiable way.

Probability is also critically useful for reasoning and learning from data. Probabilistic reasoning is beyond the scope of this book, but be assured that it is built upon the understanding of probability theory provided in this chapter. Fortunately, the applications of probabilistic reasoning are easily understood. Probabilistic reasoning is largely responsible for the recommendations that Amazon.com and Netflix provide to their users. To some degree, both companies match each user to a set of other users with similar preferences (based on data the companies possess) and then utilize the preferences of these similar users to generate the recommendations. But the matches between users are not exact—they are probabilistic—and the suggestions based on these matches also are probabilistic. These recommender systems try to utilize all of the probabilistic information they have so as to obtain the most reliable recommendations. Similarly, probabilistic reasoning can be used to suggest friends or colleagues to add to your Facebook or LinkedIn account. Social network companies have a great incentive to make good suggestions since the more connections that are made, the more valuable the site and the more likely a user is to stay active on the site and not leave for a competitor's site. In the social networking case these suggestions are made by taking separate pieces of information, or evidence, assigning probabilities to them, and then combining these probabilities so that only the most likely friends or colleagues are suggested as possible matches.

A basic knowledge of probability is also required to understand statistics, which is an important field for anyone who wants to be able to describe, understand, and draw inferences from data. Statistics is related to probability in a very fundamental way and one cannot understand statistics without first having at least a basic understanding of probability.

7.1 Terminology and Background

Before we can discuss probability, we need to define some basic terms. These terms should help us to precisely define various probability problems, such as drawing a flush in poker or winning the state lottery. We choose to describe these as *experiments* with *outcomes*. Typically we are not interested in determining the probability of all outcomes, but instead a specific set of outcomes, which we refer to as an *event*. In the coin tossing example where we want to calculate the probability of tossing a head, the event is "tossing a head." Note that an event is a specific *set* of outcomes because for many problems we are interested in the probability of any one of a number of outcomes occurring. For example, we may be interested in the probability of selecting a 10 or an Ace from a deck of cards.

Some notation is needed so that we can concisely describe and reason about these probabilistic statements. We have already defined an event; by convention we use E to represent an event (or E_1 and E_2 if there are two events). The set of all possible outcomes is denoted by S, which we also call the *sample space* of E. Note that both E and S have cardinalities associated with them, denoted by $|E|$ and $|S|$. Our goal is to calculate the probability that the observed outcome belongs to the set E. We denote this probability as $\text{Prob}(E)$.

Before providing the formula to calculate $\text{Prob}(E)$, let us first apply this terminology to an example from the domain of dice, where we are interested in predicting the probability of rolling a 7 on two dice. Assuming that $(3, 2)$ represents rolling a 3 on the first die and a 2 on the second die, we would then describe the probability problem as determining the value of $\text{Prob}(E)$, where

$$\begin{aligned} E &= \{(1,6),(2,5),(3,4),(4,3),(5,2),(6,1)\} \\ S &= \{(1,1),(1,2),\ldots,(1,6), \\ &\quad (2,1),(2,2),\ldots,(2,6), \\ &\quad (3,1),(3,2),\ldots,(3,6), \\ &\quad \vdots \\ &\quad (6,1),(6,2),\ldots,(6,6)\}. \end{aligned}$$

The formula for calculating $\text{Prob}(E)$ is quite simple:

$$\text{Prob}(E) = \frac{|E|}{|S|} \tag{7.1}$$

This formula should make intuitive sense: the probability of observing the desired event is simply the fraction of the sample space covered by the event. If we apply equation (7.1) to the dice problem, we get:

$$\text{Prob}(E) = \frac{|E|}{|S|} = \frac{6}{36} = \frac{1}{6} = 0.16667\,.$$

Note that in this example we determine the size of the sets S and E by enumerating the elements belonging to the set. This is feasible here because the sets are relatively small. In many cases this will not be possible and we will need to use the techniques we learned from the previous chapter on counting. For example, in this case rather than explicitly enumerating the elements in S we could have determined $|S|$ using the multiplication rule for counting: there are 6 outcomes for the first toss and 6 outcomes for the second toss, so there are $6 \times 6 = 36$ total possible outcomes and hence $|S| = 36$.

Note that the probability is a decimal value and that the probability of any outcome will always be between 0 and 1, inclusive. This is because the value of $|E|$ can vary from 0 to $|S|$. A probability of 0 means that the event can never occur and a value of 1 indicates that the event will always occur. As an example of these two extremes, the probability of rolling two six-sided dice and having them sum to 15 is 0, while the probability of them having a sum greater than 0 is 1. If you ever compute a probability that is greater than 1 or less than 0 or a probability that is a negative number, then you know that an error was made and you should recheck your calculations. For classwork and on exams it may be acceptable to leave the probability expressed as a fraction, depending on the policy of your instructor and whether a calculator is handy. Finally, it is generally acceptable to express probabilities as percentages and, at least in informal settings, this is often the most natural alternative. For example, one might say in conversation, "If I flip this coin I have a 50% chance of getting a head."

While equation (7.1) for calculating the probability of an event is simple and should make intuitive sense, there are two subtle assumptions built into this equation:

1. The sample space is *finite*.

2. All observations are *equally likely*.

These assumptions hold for a wide class of problems, including virtually all that we study in this chapter, but does not cover all problems. For example, when rolling a fair die all six values are equally likely; when drawing a card from a deck of cards all fifty-two cards have equal likelihood of being drawn. However, the second assumption does not hold for a "weighted" die. As another example, if we have a bag of balls and reach in to pick a ball, the probability of selecting each ball is not equal if some balls are larger than others. Toward the end of the chapter we will learn to deal with some cases where all outcomes are not equally likely.

In the remainder of this section we provide some very simple probability examples. Any complexity for these basic examples comes from the need to "count" the number of elements in the sets E and S. For these examples we define E and S very carefully. Once one gains experience with probability, it may not be necessary to explicitly define these values—at least for very simple problems. For example, if you are asked "What is the probability of guessing a number between 1 and 10?" it is reasonable to answer 0.1 without any further explanation. In some cases there will be too many elements in the sets S or E to enumerate. Since we only need to the know the cardinality of these sets, in these cases one can just specify the cardinality of these sets.

Example 7.1: Determine the probability that you flip a coin three times and that you get all heads.

Solution: We represent the three coin flips using a triple, such that (T,H,T) denotes the case where the first and third flips yield a tail and the second flip yields a head. Then we can compute the desired probability as:

$$\begin{aligned} S &= \{(\text{H,H,H}), (\text{H,H,T}), (\text{H,T,H}), (\text{H,T,T}), (\text{T,H,H}), (\text{T,H,T}), \\ &\quad (\text{T,T,H}), (\text{T,T,T})\} \\ E &= \{(\text{H,H,H})\} \\ \text{Prob}(E) &= \frac{|E|}{|S|} = \frac{1}{8} = 0.125\,. \end{aligned}$$ □

Example 7.2: Determine the probability that you flip a coin 10 times and that you get all heads.

Solution: In the previous example it was feasible to enumerate the size of the sets S and E, so the cardinality of the two sets were determined by explicitly enumerating them. In this example that is not feasible for S, due to its size, so we use the multiplication rule for counting. While the set E could be enumerated, we also use the multiplication rule for counting to determine its size. We find:

$$\begin{aligned} |S| &= 2 \cdot 2 \cdot 2 \cdot 2 \cdot 2 \cdot 2 \cdot 2 \cdot 2 \cdot 2 \cdot 2 = 2^{10} = 1{,}024 \\ |E| &= 1 \cdot 1 \cdot 1 \cdot 1 \cdot 1 \cdot 1 \cdot 1 \cdot 1 \cdot 1 \cdot 1 = 1^{10} = 1 \\ \text{Prob}(E) &= \frac{|E|}{|S|} = \frac{1}{1{,}024} = 0.00098. \end{aligned}$$ □

Example 7.3: Determine the probability that you flip a coin three times and that you get exactly two heads.

Solution: Using the same notation as in Example 7.1, we can compute the desired probability as:

$$S = \{(\text{H,H,H}), (\text{H,H,T}), (\text{H,T,H}), (\text{H,T,T}), (\text{T,H,H}), (\text{T,H,T}), (\text{T,T,H}), (\text{T,T,T})\}$$

$$E = \{(\text{H,H,T}), (\text{H,T,H}), (\text{T,H,H})\}$$

$$\text{Prob}(E) = \frac{|E|}{|S|} = \frac{3}{8} = 0.375\,. \qquad \square$$

Example 7.4: Given a standard deck of cards, which is the probability of drawing one card and that card being a 2 or a 3?

Solution: We formulate the problem as follows:

$$|S| = 52,$$

$$E = \{2\clubsuit, 2\diamondsuit, 2\heartsuit, 2\spadesuit, 3\clubsuit, 3\diamondsuit, 3\heartsuit, 3\spadesuit\}, \text{ so that } |E| = 8,$$

$$\text{Prob}(E) = \frac{8}{52} = \frac{2}{13} = 0.154. \qquad \square$$

7.2 Complement

As already discussed, an event is a set of outcomes from the sample space S. We can also talk about the *complement* of any set of outcomes, but we will usually be interested in the complement of a specific event E, which is typically denoted by E' or $\overline{E}$; we use the E' symbol in this text. The complement E' of E is the event consisting of all the outcomes that do *not* occur in E; equivalently, E' consists of all the outcomes in $S - E$. Note that our use of the $'$ symbol for the complement of an event is consistent with our use of $'$ for the complement of a set in Chapter 1, which is a good thing, since an event is merely a special kind of set.

An example will help make this clear. The complement of flipping a coin and getting a head is flipping a coin and getting a tail and the complement of "drawing a blackjack" is "not drawing a blackjack." Note that in the second case we describe the complement as what is not drawn only because there are too many items in $S - E$ to enumerate.

What can we say about the probability of E', denoted Prob(E'), if we know Prob(E)? First note that $E \cup E' = S$, given our definition of complement. Since we already established that the probability that an outcome in S occurs is 1, we know that $\text{Prob}(E) + \text{Prob}(E') = \text{Prob}(S) = 1$. Solving for for E' we get:

$$\text{Prob}(E') = 1 - \text{Prob}(E). \tag{7.2}$$

Sometimes it is difficult to calculate the probability of an event but much easier to calculate the probability of its complement. When this is the case, one should just calculate the probability of the complement and then use the formula above to calculate the required probability. This is demonstrated in the example below.

Example 7.5: Given a class with eight students, what is the probability that at least two students will share the same birth month?

Solution: We begin as usual, by formalizing the problem, letting

$$S = \text{“all possible assignments of months to the eight students,”}$$
$$E = \text{“assignments of months to students,}$$
$$\text{where at least one month is repeated.”}$$

The value of $|S|$ is easy to calculate. Since birth months can be repeated and there are 12 choices for each of the 8 students, we find that $|S| = 12^8 = 429{,}981{,}696$. However, it is not easy to calculate $|E|$ directly. One approach is to break it into cases, but since a month can be repeated between two and eight times, there would be many cases to evaluate. The problem becomes much easier once we recognize that:

E' = assignments of month to students in which all months are different.

In this case, E' can be computed using the techniques from the chapter on counting. There are 12 choices for the first student, then 11 for the second, etc. We get

$$|E'| = P(12, 8) = 12 \cdot 11 \cdot 10 \cdot 9 \cdot 8 \cdot 7 \cdot 6 \cdot 5 = 19{,}985{,}400.$$

Thus

$$\text{Prob}(E') = \frac{|E'|}{|S|} = \frac{19{,}985{,}400}{429{,}981{,}696} = 0.0465 = 4.65\%$$

Applying equation (7.2), we get

$$\text{Prob}(E) = 1 - \text{Prob}(E') = 1 - 0.0465 = 0.9535 \text{ or } 95.35\%$$

Thus we see that it is very likely that if we have eight students in a class that a birth month will be repeated amongst these students. □

7.3 Elementary Rules for Probability

In this section we introduce two basic rules for combining probabilities, just as we did for counting in the previous chapter: the addition rule and the multiplication rule. These rules are then examined in more depth through the use of examples. However, these elementary rules are only applicable when certain conditions hold; otherwise the more general (and slightly more complex) versions of these rules, which are described in Section 7.4, must be used. Before we can discuss these conditions, two additional terms must be introduced.

Our first concept is *disjointness.* We say that two or more events are *disjoint* if the outcomes associated with one event are not present in the outcomes of any of the other events (i.e., if the events form non-overlapping sets). More formally, two events E_1 and E_2 are disjoint if $E_1 \cap E_2 = \emptyset$.

Our second concept is *independence.* Informally, we say that two events E_1 and E_2 are *independent* if the outcome of any one of these events does not *in any way* impact or influence the outcome of the other event. This informal definition can lead to some ambiguity. So, as is common in mathematics, we also provide a more formal definition, which in some cases will make it easier to determine whether two events are independent. Namely, we will say that the events E_1 and E_2 are *independent* if

$$\text{Prob}(E_1 \cap E_2) = \text{Prob}(E_1) \cdot \text{Prob}(E_2). \tag{7.3}$$

Example 7.6: A six-sided die is rolled. Are the events "roll an odd number" and "roll an even number" disjoint? Are they independent?

Solution: Let

$$E_1 = \text{"roll an odd number"} = \{1, 3, 5\}$$

and

$$E_2 = \text{"roll an even number"} = \{2, 4, 6\}.$$

Since $E_1 \cap E_2 = \emptyset$ contains no elements, the events are disjoint. However, the events are not independent. Using the informal criterion, if we know that event E_1 has occurred, then event E_2 can *never* occur. Moreover, if E_1 has not occurred, then E_2 *must* occur. Since the outcome of E_1 influences the outcome of E_2, the events E_1 and E_2 are not independent.

Now let's use the more formal criterion. The events are independent if and only if (7.3) holds. Based on our knowledge of probability, we know that $\text{Prob}(E_1) = \frac{3}{6}$ and $\text{Prob}(E_2) = \frac{3}{6}$. Hence $\text{Prob}(E_1) \cdot \text{Prob}(E_2) =$

$\frac{9}{36} = \frac{1}{4}$. We now compute Prob$(E_1 \cap E_2)$, which we can do directly. This is especially easy for this example; since $E_1 \cap E_2 = \emptyset$, we see that Prob$(E_1 \cap E_2) = 0$. So (7.3) does *not* hold, and thus the events are not independent. □

The second non-independence proof in this last example actually shows us a useful general result: *disjoint events (having non-zero probabilities) are never independent.*

Example 7.7: [SN] Are the following pairs of events independent?

- E_1: Manny and Marina are friends, live in the same city, and both list their social status as married.
- E_2: Manny and Marina are married to one another.

Solution: Events E_1 and E_2 are not independent. While the facts represented in E_1 (taken from our social network example) do not require E_2 to be true, they influence the likelihood that E_2 occurs. To see this, imagine that we take all pairs of people in large social network that are friends *and* live in the same city *and* both list their status as married. The fraction of these pairs of people that are actually married can be interpreted as an estimate of the probability of two people being married given that they satisfy the three conditions (friends, same city, and both married). Most people would agree that this probability should be much greater than the probability that two random people selected from the social network are married to one another. Of course, this hypothesis could be experimentally verified from the data, although this still would not "prove" the hypothesis. □

Remark. In this example the probabilistic reasoning process was based on assumptions that are not formally justified. But in actuality a great deal of real-world probabilistic reasoning is based on assumptions that are often unproven. In fact, statistics, which is based on probability theory, relies on assumptions about the data that in practice are frequently *known* not to hold! But we often find that even if these assumptions do not hold perfectly, we can still obtain reasonable results. □

The elementary addition rule for probability is only applicable when the events have disjoint outcomes and the elementary multiplication rule for probability is only applicable when the events are independent. In Section 7.4 we introduce the general versions of these rules and demonstrate how to apply them when these assumptions do not hold.

7.3.1 The Elementary Addition Rule

The elementary addition rule for probability, described below, can be used to combine the probabilities for disjoint events.

Addition Rule (elementary version)

Given two disjoint events, E_1 and E_2, with associated probabilities $\text{Prob}(E_1)$ and $\text{Prob}(E_2)$, the probability of E_1 *or* E_2 occurring is equal to $\text{Prob}(E_1) + \text{Prob}(E_2)$. That is,

$$\text{Prob}(E_1 \cup E_2) = \text{Prob}(E_1) + \text{Prob}(E_2)$$

if E_1 and E_2 are disjoint.

We can justify this formula as follows. Since E_1 and E_2 are disjoint, from our knowledge of sets from Chapter 1, we know that

$$|E_1 \cup E_2| = |E_1| + |E_2|$$

So

$$\begin{aligned}\text{Prob}(E_1 \cup E_2) &= \frac{|E_1 \cup E_2|}{|S|} = \frac{|E_1|}{|S|} + \frac{|E_2|}{|S|} \\ &= \text{Prob}(E_1) + \text{Prob}(E_2).\end{aligned}$$

Many problems can be solved without using the addition rule for probability, simply by using an alternative formulation of the problem. Whether you need the addition rule depends on whether you formulate a problem as having a single event, in which case you do not need it, or if you have multiple events. The example below illustrates this.

Example 7.8: What is the probability of drawing a blackjack in the initial deal from a fresh deck? Reminder: a blackjack occurs if the two cards dealt consist of an ace and a card with a numerical value of 10 (i.e., 10, Jack, Queen, or King).

Solution: One thing that must be realized is that a blackjack occurs by drawing an Ace followed by a card with the value of 10 *or* by drawing a card with value 10 followed by an Ace. This means that the addition rule of probability applies and that we should calculate the component probabilities and then add them. We do not show the complete formulation for the sub-problems since you should be familiar with them by now.

We want Prob(blackjack), but from the addition rule of probability we know that

$$\text{Prob(blackjack)} = \text{Prob(ace then value=10)} + \text{Prob(value=10 then ace)}.$$

Note that there are 4 aces in a deck and 16 cards with a numerical value of 10. Thus

$$\begin{aligned}
&\text{Prob(ace then value=10)} \\
&\quad = \text{Prob(ace on first draw)} \cdot \text{Prob(value=10 on second draw)} \\
&\quad = \frac{4}{52} \cdot \frac{16}{51} = 0.024
\end{aligned}$$

and

$$\begin{aligned}
&\text{Prob(value=10, ace)} \\
&\quad = \text{Prob(value=10 on first draw)} \cdot \text{Prob(ace on second draw)} \\
&\quad = \frac{16}{52} \cdot \frac{4}{51} = 0.024\,.
\end{aligned}$$

Thus Prob(blackjack) $= 0.024 + 0.024 = 0.048$ (or 4.8%).

Note that we can use the elementary addition rule for probability because the two events are disjoint. They are disjoint because you can never draw an Ace on the first draw and card with value of 10 on the second draw *and* also draw a card with value 10 on the first draw and an ace on the second draw. □

Example 7.9: A person rolls a die and wins a prize if the roll is a 1 or a 2. Using the elementary addition rule for probability, what is the probability of winning?

Solution: It is trivial to solve this without using the addition rule, since we could just have $E = \{1, 2\}$ and $S = \{1, 2, 3, 4, 5, 6\}$ and therefore deduce that the probability is $\frac{2}{6}$. But we can also break this problem into two subproblems and then solve it using the addition rule as follows (using the second formulation of the addition rule). We find

$$\begin{aligned}
E_1 &= \{1\}, \\
E_2 &= \{2\},
\end{aligned}$$

$$\text{Prob(1 or 2)} = \text{Prob}(E_1) + \text{Prob}(E_2) = \frac{1}{6} + \frac{1}{6} = \frac{1}{3} = 0.333\,.$$

The elementary addition rule of probability holds here because E_1 and E_2 are disjoint, since one cannot get a 1 and a 2 on the same roll of the die. In this example we used the addition rule, but since the problem is so simple one could easily have computed the solution directly. However, for more complex problems this will not be the case.

7.3.2 The Elementary Multiplication Rule

The elementary multiplication rule for probability can be used if all events are independent. The rule is described formally below.

> **Multiplication Rule (elementary version)**
>
> Given two independent events, E_1 and E_2, with associated probabilities $\text{Prob}(E_1)$ and $\text{Prob}(E_2)$, the probability of E_1 and E_2 occurring, $\text{Prob}(E_1 \text{ and } E_2)$ is equal to $\text{Prob}(E_1) \cdot \text{Prob}(E_2)$. That is, $\text{Prob}(E_1 \cap E_2) = \text{Prob}(E_1) \cdot \text{Prob}(E_2)$.

The elementary version of the multiplication rule is just a restatement of the definition of independent events provided earlier in this chapter in equation (7.3). The examples that follow will hopefully illustrate that this rule agrees with your intuition about how probabilities should be combined.

Many problems can be solved without using the multiplication rule for probability, simply by using an alternative formulation of the problem. Whether you need the multiplication rule depends on whether you formulate a problem as having a single event, in which case you do not need it, or having multiple events. The first example below will illustrate this.

Example 7.10: In Example 7.1 we calculated the probability of flipping a coin three times and getting all heads. In that example we did not use the multiplication rule. Solve that problem using the multiplication rule by defining each event as covering only a single toss of the coin.

Solution: We formulate the problem as follows. Our constituent events are

$$\begin{aligned} E_1 &= \text{“first toss is a head”}, \\ E_2 &= \text{“second toss is a head”}, \\ E_3 &= \text{“third toss is a head”}, \end{aligned}$$

and so we have

$$\text{Prob}(3 \text{ heads}) = \text{Prob}(E_1) \cdot \text{Prob}(E_2) \cdot \text{Prob}(E_3)\,.$$

We now have to calculate the three component probabilities. The probability of getting a head for each toss is $\frac{1}{2}$ (since $|E_1|$, $|E_2|$ and $|E_3|$ are all 1 and $|S|$ is 2 for all three events). Thus our solution is

$$\text{Prob}(3 \text{ heads}) = \frac{1}{2} \cdot \frac{1}{2} \cdot \frac{1}{2} = \frac{1}{8} = 0.125$$

In Example 7.1 we solved this using enumeration. Since the enumeration method is not always feasible, we should compare the current method, which uses the multiplication rule for probability, to the multiplication rule for counting. Using the multiplication rule for counting, in the manner it was used in Example 7.2 (but for three heads instead of ten), we get:

$$\text{Prob}(3 \text{ heads}) = \frac{1 \cdot 1 \cdot 1}{2 \cdot 2 \cdot 2} = \frac{1}{8} = 0.125.$$

Clearly these formulas are equivalent in that they yield the same result. The only difference is that the two methods have you think about the problem in slightly different ways. In the approach taken in Example 7.2, we view the multiple coin tosses as one event. This requires a bit more work when doing the counting. Using the current method, the counting is simpler, but we need to deal with more events. You should be able to use either method, but which one is more natural depends on how you think of the problem. In many cases the multiplication rule for probability is, for most people, the most natural way of thinking of a problem. For example, if you ask most people, especially those who have not studied probability recently, what the probability of getting two heads is in coin tosses, they will probably reason: the probability of each is 0.5 so the probability of both is $0.5 \cdot 0.5 = 0.25$. □

Example 7.11: You have decided on a specific model of car to buy but are having a problem deciding on a few of the options. The options include the color (which has eight choices including red), whether to have air conditioning, and whether to have the 4-wheel drive option. Finally, in frustration, you decide to select each option randomly. What is the probability that you will wind up with a red car, with air-conditioning, but without the 4-wheel drive option? Use the multiplication rule of probability (not of counting) to calculate the answer.

Solution: Our constituent events are

$$E_1 = \text{"choose red color"},$$
$$E_2 = \text{"choose air-conditioning (A/C) option"},$$
$$E_3 = \text{"choose 4-wheel drive option"}.$$

Each event has a different sample space. For this problem the sizes of the corresponding sample spaces are: $S_1 = 8$, $S_2 = 2$, and $S_3 = 2$. All of the three events have a cardinality of exactly 1. Thus we get

$$\begin{aligned}\text{Prob(red, A/C, no 4-wheel)} &= \text{Prob(red)} \cdot \text{Prob(A/C)} \cdot \text{Prob(no 4-wheel)}\\ &= \frac{1}{8} \cdot \frac{1}{2} \cdot \frac{1}{2} = \frac{1}{32} = 0.031\,. \qquad \square\end{aligned}$$

Example 7.12: [SN] Imagine that an analysis of the data from an expanded version of the social network example in the Appendix shows that if we randomly select two members, the probability that both were born in the same year is 0.040, grew up in the same town (hometown) is 0.005, currently live in the same city is 0.007, and graduated from the same high school class is 0.00022.

(a) Assuming that the events are independent, what is the probability that two randomly selected members of the social network grew up in the same hometown and currently live in the same city?

(b) Again using the independence assumption, what is the probability that two randomly selected members grew up in the same hometown and were born in the same year? Comment on the relationship between this calculated value and the given probability that two random members graduate from the same high school class.

(c) Explain why this independence assumption most likely does not hold between "same hometown" and "same current city." Would the actual probability of both events occurring be higher or lower than the calculated value?

Solution:

(a) Our constituent events are

$$\begin{aligned}E_1 &= \text{"both members grew up in same hometown"},\\ E_2 &= \text{"both members currently live in the same city"}.\end{aligned}$$

Given the independence assumption, we have

$$P(E_1 \cap E_2) = P(E_1) \cdot P(E_2) = 0.005 \cdot 0.007 = 0.000035.$$

(b) Our constituent events are

$$E_1 = \text{"both members grew up in same hometown"},$$
$$E_2 = \text{"both members have same birth year"}.$$
$$P(E_1 \cap E_2) = P(E_1) \cdot P(E_2) = 0.005 \cdot 0.040 = 0.0002.$$

The calculated value of 0.0002 is very close to the given probability of graduating from the same high school class, which is 0.00022. This is to be expected since if you grow up in the same hometown and are born in the same year, you typically will graduate in the same high school class. In this case one would expect the two constituent events to be nearly independent—but not completely so since different hometowns may have different age profiles (i.e., some towns attract retirees while others attract young couples).

(c) You are much more likely to currently live in your hometown, or a town geographically close to your hometown, than a randomly selected city. Thus the independence assumption should not hold and, furthermore, the probability that you grew up in the same hometown and currently live in the same city should be much higher than the value calculated using the independence assumption. □

7.4 General Rules for Probability

The elementary rules for probability hold in only some cases—the addition rule only holds if the events are disjoint and the multiplication rule only holds if the events are independent. If this is not the case, then the rules described in this section should be used. However, sometimes you may avoid this by exhaustively enumerating the outcomes and "manually" adjusting for the fact that the outcomes are not independent or disjoint. While this may sound complicated, it is not, and most people can solve these problems using their intuition alone. Consistent with our approach to previous probability problems, we begin with an example that can be solved directly, using intuition alone.

Example 7.13: Given a standard deck of cards, what is the probability of drawing a red card or a 2?

Solution: The size of the sample space, $|S| = 52$, since there are 52 cards in a deck. In a standard deck there are also twenty-six red cards and four 2's. The key insight is to realize that there are two red 2's and two black 2's and that if the red cards are counted then the red 2's should not be counted separately, since then they would be counted twice. Thus, the number of cards that are either red or a 2 equals $26 + 2 = 28$. So we see that

$$\text{Prob(card is red or a 2)} = \frac{28}{52} = 0.538\,. \quad \square$$

7.4.1 The General Addition Rule

The general addition rule for probability must be used when the events involved in the probability calculation are not disjoint.

Addition rule (general version)

Given events E_1 and E_2, with associated probabilities $\text{Prob}(E_1)$ and $\text{Prob}(E_2)$, the probability of E_1 *or* E_2 occurring is

$$\text{Prob}(E_1 \cup E_2) = \text{Prob}(E_1) + \text{Prob}(E_2) - \text{Prob}(E_1 \cap E_2)$$

Note that if the events are disjoint, then $E_1 \cap E_2$ will equal $\emptyset$ and therefore $\text{Prob}(E_1 \cap E_2)$ will equal 0, which means that the general addition rule will reduce to the elementary addition rule for probability. Now we can return to Example 7.13 and apply the general addition rule for probability.

Example 7.14: Solve Example 7.13 using the general addition rule for probability. That problem asked what the probability is of drawing one card and having it be either a red card or a 2.
Solution: Let $E_1 =$ "pick a red card" and let $E_2 =$ "pick a 2". Our task is to calculate $\text{Prob}(E_1 \cup E_2)$. First we must decide whether we can use the elementary addition rule of probability which means we must decide if E_1 and E_2 are disjoint. Since these two sets overlap (due to the existence of red 2's) the outcomes are not disjoint so we must use the general addition rule. We start off with the formula for that rule:

$$\text{Prob}(E_1 \cup E_2) = \text{Prob}(E_1) + \text{Prob}(E_2) - \text{Prob}(E_1 \cap E_2)$$

We now need to determine $\text{Prob}(E_1)$, $\text{Prob}(E_2)$, and $\text{Prob}(E_1 \cap E_2)$. The first two values are easy to determine. Since the size of the sample space, $|S| = 52$, and since there are twenty-six red cards and four 2's, we know that $\text{Prob}(E_1) = \frac{26}{52}$ and $\text{Prob}(E_2) = \frac{4}{52}$. The only thing new in this problem is to compute the intersection of the two events. Using a basic understanding of sets and the intersection operator, we know that the set $E_1 \cap E_2$

equals the red 2's and that there are two of them in a deck. Therefore $\text{Prob}(E_1 \cap E_2) = \frac{2}{52}$. Substituting these three values into the addition rule we get:

$$\text{Prob}(E_1 \cup E_2) = \frac{26}{52} + \frac{4}{52} - \frac{2}{52} = \frac{28}{52} = 0.538$$

Note that, as expected, this answer matches the one we computed directly in Example 7.13. □

Example 7.15: I flip a coin and roll a six-sided die. What is the probability that I toss a head or roll either a 1 or a 2?

Solution: We find

$$\begin{aligned}
S_1 &= \{\text{head, tails}\}, \text{ and so } |S_1| = 2, \\
S_2 &= \{1, 2, 3, 4, 5, 6\}, \text{ and so } |S_2| = 6, \\
E_1 &= \text{“flip coin and get head”, and so } |E_1| = 1, \\
E_2 &= \text{“roll die and get 1 or 2”, and so } |E_2| = 2.
\end{aligned}$$

Thus

$$\begin{aligned}
\text{Prob}(E_1) &= \frac{|E_1|}{|S_1|} = \frac{1}{2} \\
\text{Prob}(E_2) &= \frac{|E_2|}{|S_2|} = \frac{2}{6} = \frac{1}{3}.
\end{aligned}$$

The only step left is to compute $\text{Prob}(E_1 \cap E_2)$, the probability of getting a head and either a 1 or a 2. To do this we need to use the multiplication rule for probability. But we can only use the elementary rule version if the two events are independent. In this case they are independent since what happens on the flip of the coin has no impact on the roll of the die. Thus we can compute this simply as

$$\text{Prob}(E_1 \cap E_2) = \text{Prob}(E_1) \cdot \text{Prob}(E_2) = \frac{1}{2} \cdot \frac{1}{3} = \frac{1}{6}.$$

Now we can substitute he three computed values into the general addition rule, getting

$$\begin{aligned}
\text{Prob}(E_1 \cup E_2) &= \text{Prob}(E_1) + \text{Prob}(E_2) - \text{Prob}(E_1 \cap E_2) \\
&= \frac{1}{2} + \frac{1}{3} - \frac{1}{6} = \frac{2}{3} \\
&= 0.667.
\end{aligned}$$

□

7.4.2 The General Multiplication

The general multiplication rule for probability must be used when the events are not independent. Before we can introduce the general rule, we first must introduce the notion of *conditional probability.*

Given two events E_1 and E_2, the conditional probability of E_1 occurring given that E_2 occurs is denoted as $\text{Prob}(E_1|E_2)$. If the events are independent then $\text{Prob}(E_1|E_2) = \text{Prob}(E_1)$, since by the definition of independence E_2 has no influence on E_1. The following example will clarify the notion of conditional probability.

Example 7.16: A standard six-sided die is rolled by your friend, but in such a way that you cannot see what value comes up. Your friend tells you that the value that comes up is odd, but does not tell you what specific value was rolled. What is the probability that she rolled a 3?

Solution: Let E_1 = "a 3 is rolled". The probability of E_1 occurring without the additional knowledge that the roll was odd would be:

$$\text{Prob}(E_1) = \frac{|E_1|}{|S|} = \frac{1}{6}$$

However, we additionally know that the roll was odd. If we let E_2 = "the roll is odd", we see that $E_2 = \{1, 3, 5\}$. We are now interested in $\text{Prob}(E_1|E_2)$. Since one of the three elements of E_2 is 3, we now see that $\text{Prob}(E_1|E_2) = \frac{1}{3}$. □

The conditional probability can be determined using the equation

$$\text{Prob}(E_1|E_2) = \frac{\text{Prob}(E_1 \cap E_2)}{\text{Prob}(E_2)}. \tag{7.4}$$

This equation can be justified intuitively as follows. Imagine that the sets E_1 and E_2 are represented by intersecting ovals in a Venn diagram, where the size of each oval is proportional to the likelihood of each outcome. When we ask about the probability of E_1 given E_2, we know that E_2 has occurred and therefore that the outcome is somewhere within the oval representing E_2. The probability of E_1 then occurring is simply the fraction of E_2 that is covered by E_1, which is simply $|E_1 \cap E_2|/|E_2|$. Equation (7.4) can be rearranged using simple algebra so that that $\text{Prob}(E_1 \cap E_2)$ can be calculated, which leads to the general multiplication rule of probability. Note that $\text{Prob}(E_1 \cap E_2)$ is the same as $\text{Prob}(E_1 \text{ and } E_2)$.

Multiplication rule (general version)

Given two events E_1 and E_2, with associated probabilities $\text{Prob}(E_1)$ and $\text{Prob}(E_2)$, the probability of E_1 occurring *and* E_2 occurring is:

$$\begin{aligned}\text{Prob}(E_1 \text{ and } E_2) &= \text{Prob}(E_1) \cdot \text{Prob}(E_2|E_1) \\ &= \text{Prob}(E_2) \cdot \text{Prob}(E_1|E_2)\end{aligned}$$

We begin with a simple example that demonstrates that we have already been using the general multiplication rule for probability without even knowing it.

Example 7.17: What is the probability of picking two aces from a deck of cards?

Solution: Let E_1 = "pick Ace on first draw" and E_2 = "pick Ace on second draw." We then want to determine $\text{Prob}(E_1 \text{ and } E_2)$. We only need to use the general rule if the events are not independent. Are they independent? In this case they are not independent since once we draw the first card it changes the deck and that impacts the second draw. Specifically, if we pick an Ace on the first draw then the probability of picking an Ace on the second draw goes down substantially, whereas if we do not pick an Ace on the first draw then the probability of picking an Ace on the second draw goes up slightly. Applying the general rule we get

$$\begin{aligned}\text{Prob}(E_1 \text{ and } E_2) &= \text{Prob}(E_1) \cdot \text{Prob}(E_2|E_1) \\ &= \frac{4}{52} \cdot \frac{3}{51} = 0.0045\,.\end{aligned}$$ □

Note that we solved problems like this even before we learned about conditional probability. However, now we have the terminology and background to better understand the complexities involved in the problem.

7.5 Bernoulli Trials and Probability Distributions

In this section we begin by discussing Bernoulli trials and then introduce probability distributions. A Bernoulli trial is an experiment where there are only two possible outcomes. The outcomes might be "yes" or "no" or perhaps "0" or "1". The probability of one of the outcomes is referred to as p, so that probability of the other outcome is $1 - p$. An excellent example of a Bernoulli trial is flipping a coin and recording whether it lands with the "head" or "tail" side facing up. Currently we know how to compute various probabilities related to Bernoulli trials, such as "What is the probability of flipping three coins and getting no heads?" However, we have not yet tried to answer more complex probability questions, such as "What is the probability of flipping a coin five times and getting exactly three heads?" There is a formula for computing the answer to these types of questions. However, this formula is a bit complex and it is often easier to compute the solution from first principles. To illustrate this, we begin with a simple example, which we solve directly from first principles.

Example 7.18: What is the probability of tossing a coin five times and getting exactly three heads?

Solution: There are many different ways that we can get exactly three heads. However, in this case we do not care where the heads come up in the sequence— we only care that there are three of them. We can therefore solve the problem by first determining the probability of getting three heads for one specific permutation and then multiply this value by the number of permutations that yield three heads.

We can start by picking any permutation that has exactly three heads, so we arbitrarily start with the permutation (H,H,H,T,T). What is the probability of getting this specific permutation? Since the probability of getting an H is $\frac{1}{2}$ and the probability of getting a T is $\frac{1}{2}$, the probability of getting this specific permutation is simply:

$$\frac{1}{2} \cdot \frac{1}{2} \cdot \frac{1}{2} \cdot \frac{1}{2} \cdot \frac{1}{2} = \frac{1}{32} = 0.03125$$

Next we need to determine how many permutations will yield exactly three heads. Since there are five positions and three H's, we can place the three H's in the five positions in $C(5, 3)$ ways. Thus the answer to the problem is $C(5, 3) \cdot 0.03125 = 10 \cdot 0.03125 = 0.3125$. □

We can generalize the solution to Example 7.18 as follows. Let K be the actual number of heads that appear if we perform n experiments, each with a probability p of obtaining the desired outcome. The probability of obtaining exactly k of the desired outcomes is then given by

$$\text{Prob}(K = k) = C(n, k)p^k(1 - p)^{n-k} . \qquad (7.5)$$

Let's review this equation to see how it is a generalization of the solution for Example 7.18. The left side of the equation simply says that we are computing the probability of obtaining exactly k of the desired outcomes. If we want to compute this probability for exactly 3 of the desired outcomes then we would compute $\text{Prob}(K = 3)$. Note that k is a variable because the formula needs to work for a variety of values. Given that there are a total of n experiments and only two possible outcomes, since we need exactly k of the desired outcomes we must also have $n-k$ of the other outcomes. Since the probabilities of the two outcomes of a Bernoulli trial must sum to 1, we see that the probability of getting one specific permutation is $p^k(1-p)^{n-k}$, where p^k is the probability of getting k of the desired outcomes and $(1-p)^{n-k}$ is the probability of getting the other outcomes. Finally, there are $C(n,k)$ ways to place k items into n places. Hopefully you will agree that it is not very difficult to reproduce this reasoning; in fact, reproducing this reasoning may be easier than memorizing equation (7.5).

Equation (7.5) is called the *binomial distribution*. It is an example of a *probability distribution* because it specifies the probabilities of all possible outcomes over the sample space.

Example 7.19: Specify the probability distribution associated with flipping a coin.

Solution: We specify the probability distribution by determining the probability of all outcomes in S. For this example $S = \{\text{head, tail}\}$, so we only need to determine these two probabilities. Thus, the probability distribution is fully specified by: Prob(head) $= 0.5$ and Prob(tail) $= 0.5$. □

Example 7.20: Calculate the probability distribution associated with tossing a coin five times, as was done in Example 7.18.

Solution: Calculating the probability distribution means determining the probability of all possible outcomes. Consistent with Example 7.18 we express the outcomes with respect to the number of heads that are observed. To determine the probabilities, we use the same reasoning as in Example 7.18 or, alternatively, just plug the numbers into equation (7.5). The final results are shown in the table below; alternatively, they could have been displayed using a bar chart.

Number of Heads	0	1	2	3	4	5
Probability	$\frac{1}{32}$	$\frac{5}{32}$	$\frac{10}{32}$	$\frac{10}{32}$	$\frac{5}{32}$	$\frac{1}{32}$

There are some interesting things we can say about the binomial distribution. First of all, it is symmetric about its mean (i.e., average). Probability theory states that for the binomial distribution the mean will be np, which in this case is $5 \cdot 0.5 = 2.5$. The probability distribution also says something about how the outcomes are spread out. Based on the results for this example, we see that the outcomes close to the mean are much more likely than those further away from the mean. For example, you are ten times more likely to observe either two heads or three heads than to observe zero heads or five heads. Probability distributions play a key role in statistics and the knowledge of probability provided in this chapter should prove very useful for anyone who will study statistics. □

7.6 Expected Value

All non-trivial probability problems permit a variety of outcomes to occur. Sometimes we want to characterize these outcomes. One way of doing this, as we saw in Section 7.5, is to specify the likelihood of occurrence of each outcome in S using a probability distribution. However, sometimes we do not need such a detailed description and a more summarized description can be more useful. For numerical outcomes, one way to summarize the outcomes is to compute an "average" value. But what do we mean by the average value for a set of numerical outcomes? Normally, we calculate the average of a set of numerical values by summing the values and dividing by the number of values in the set. However, in this case we want to be able to determine the "average" value without actually observing any outcomes (e.g., without actually rolling a die and observing the number that comes up). We want to know what the average value would be based on the probability of each outcome occurring. Thus, outcomes that are more likely should count more, since they will tend to be observed more often. Thus, we need to weight each value by the probability that it occurs, which yields the *expected value* of the outcomes. Suppose that an event X has outcomes $x_1, x_2, \ldots, x_n$. Then the expected value $E[X]$ of X is given by

$$E[X] = \sum_{j=1}^{n} p_j x_j = p_1 x_1 + p_2 x_2 + \cdots + p_n x_n, \tag{7.6}$$

where $p_j = \text{Prob}(X = x_j)$ is the probability that $X = x_j$, i.e., the probability that the jth outcome occurs. Equation (7.6) is simply the weighted average of the numerical outcomes, where each weight is the probability associated with each outcome. The value we compute is called the expected value because if we repeat event X many times and average the resulting values, we expect to get this value.

Example 7.21: Given a standard six-sided die that is known to be fair, what is the expected value if the die is tossed once?

Solution: The outcomes associated with tossing a die once can be represented as the set $S = \{1, 2, 3, 4, 5, 6\}$. Given that the die is fair, we know that the probability of each of the six numerical outcomes is equal, which means that each outcome has a probability of $\frac{1}{6}$ of occurring. Thus, using equation (7.6) for expected value, we get

$$\begin{aligned}\text{Expected value} &= \frac{1}{6} \cdot 1 + \frac{1}{6} \cdot 2 + \frac{1}{6} \cdot 3 + \frac{1}{6} \cdot 4 + \frac{1}{6} \cdot 5 + \frac{1}{6} \cdot 6 \\ &= \frac{1}{6} + \frac{2}{6} + \frac{3}{6} + \frac{4}{6} + \frac{5}{6} + \frac{6}{6} \\ &= \frac{21}{6} = 3.5\end{aligned}$$

In this case the expected value of 3.5 also is the average value of the six possible outcomes, since the weights (i.e., probabilities) are all equal. This means that if you were to toss a fair die many times, the average observed value will tend to be near 3.5 (and will converge to exactly 3.5 given an infinite number of tosses). □

The next example is far more interesting and shows what happens where the probability of each outcome is not equal.

Example 7.22: A standard six-sided die is modified so that the probability of observing a 6 is five times that of observing any other outcome. What is the expected value if the die is tossed once?

Solution: As before, the set $S = \{1, 2, 3, 4, 5, 6\}$ can represent the outcomes. We must first compute the probability associated with each outcome. To do this we must employ some elementary algebra. Let y equal the probability of observing any outcome other than a 6. Then the probability of observing a 6 would be equal to $5y$. Since the probability of observing each individual outcome between 1 and 5 is y and the probability of observing a 6 is $5y$, the probability of observing an outcome in S is $5y + 5y = 10y$. Since the probability of all outcomes in S must sum to 1, we know that $10y = 1$ and thus $y = 0.1$. Thus the probability of observing a 6 is 0.5 and the probability of observing any other value is 0.1. We can then compute the expected value of the toss as follows:

$$\begin{aligned}\text{Expected value} &= 0.1 \cdot 1 + 0.1 \cdot 2 + 0.1 \cdot 3 + 0.1 \cdot 4 + 0.1 \cdot 5 + 0.5 \cdot 6 \\ &= 0.1 + 0.2 + 0.3 + 0.4 + 0.5 + 3 = 4.5\,.\end{aligned}$$

Note that with this weighted die the expected value is now up to 4.5. If the value of a 6 is more advantageous in a game of chance, the weighted die can be used to favor the "roller." In fact, weighted dice, also known as "loaded dice" do exist and for that reason casinos are very careful to ensure that a roller is not given the opportunity to surreptitiously switch the dice. □

Example 7.23: You play a lottery where to win you must pick six numbers and all must match the numbers selected from a bin of numbered ping-pong balls (but you do not need to pick them in the correct order). The bin has ping-pong balls numbered from 1 to 30 with no repetitions. If a lottery ticket costs \$1 and the prize is \$500,000, what is the expected value associated with your purchasing and filling out one lottery ticket?

Solution: To answer this problem, you must first compute your expected winnings and then subtract the \$1 entrance fee from this. To determine your expected winnings, you must compute the probability of winning the lottery. In order to do this, we need to determine the effective size of the sample space S. Since order does not matter, we have

$$|S| = C(30, 6) = \frac{30!}{24! \cdot 6!} = 593{,}775.$$

So the probability of winning from one lottery ticket is $\frac{1}{593{,}775}$ and therefore the expected winnings from a single lottery ticket is $\$500{,}000 \cdot \frac{1}{593{,}775} = \$0.842 = 84.2¢$. This would be the expected value associated with the lottery if the lottery were free. But since it costs \$1 to enter, the expected value is

$$\text{Expected value} = \$0.842 - \$1 = -\$0.158 = -15.8¢.$$

This means that each time you enter the lottery you can expect to lose 15.8¢. Of course most lotteries have a negative expected value, which means that the organization running the lottery generally shows a profit. Of course you will never lose 15.8¢—you will either win very big (very rarely) or lose \$1 (most of the time). But if you play an enormous number of times, you can expect to realize the average outcome of a 15.8¢ loss per lottery ticket. □

7.7 Exercises

7.7.1 In a certain lottery game, five different numbers are chosen between 1 and 49 where order does not matter. Each time a number is chosen, it is not returned to the bin (so the same number cannot be drawn twice). Then a bonus number is chosen between 1 and 42 (from a different set of lottery balls).

(a) What is the size of the sample space?

(b) What is the probability that someone who buys one ticket will win this lottery?

(c) What is the probability that the individual gets the bonus number correct, regardless of the other number the individual picked?

7.7.2 Find the probability of getting a number divisible by 3 when a standard six-sided die is rolled.

7.7.3 When two dice are thrown, find the probability that the sum of the dots is greater than or equal to 6.

7.7.4 To win a lottery, you need to guess the number that is randomly chosen. The number is between 1 and 10,000. Each lottery ticket costs $1.

(a) If the payout is $5,000, should the lottery commission expect to make money? Answer yes or no, and justify your answer.

(b) If the payout is $20,000, should the lottery commission expect to make money? Answer yes or no, and justify your answer.

(c) What is the minimum number of tickets you would you need to buy to guarantee that you win once?

7.7.5 Given one six-sided die (with sides numbered 1 through 6) and a four-sided die (with sides numbered 1 through 4), what is the probability of rolling both dice and obtaining the following outcomes?

(a) Rolling "snake-eyes" (i.e., a 1 on both dice).

(b) Having the sum of both dice be greater than 8.

(c) Rolling a bigger number on the 4-sided die.

(d) Rolling the same number on both dice.

7.7.6 You flip a fair two-sided coin ten times. What is the probability of getting at least one tail?

7.7.7 A red hat contains ten numbers $(1, 2, \ldots, 10)$ and a blue hat contains ten letters (A, B, ..., J). You draw one item from each hat and will win a prize if the number is less than or equal to 4 or the letter is an A, B, or C (or both occur). What is the probability that you win a prize?

7.7.8 You are given a coin which is known to be fair (it hasn't been modified and it isn't more likely to come up heads than tails). You flip the coin three times and it comes up heads three times. What is the probability that the fourth flip will be a head?

7.7.9 An exam has five true/false questions, followed by two multiple choice questions, each multiple-choice question having ten possible answers. Suppose that you guess on every question. What is the probability that you get a perfect grade?

7.7.10 For each of the following pairs of events, specify whether the events are independent. You need not prove or justify your answer.

(a) E_1: Roll a fair 6-sided die and get a 1.

E_2: Roll the same die again and also get a 1.

(b) E_1: Roll a loaded (unfair) 6-sided die and get a 6.

E_2: Roll the same die again and get a 6.

(c) E_1: A card is drawn from a freshly mixed deck of cards and is not placed back into the deck.

E_2: A second card is drawn from the deck.

7.7.11 **[SN]** Specify whether the following pairs of events are independent and justify your answer. State any relevant assumptions. These questions are based on the sample social network on page 317 and the tables of social network data that follow it. However, these questions can be answered without referring to that data.

(a) E_1: Kyle and Stan are friends.

E_2: Kyle and Stan both live in the same city (South Park) and grew up in the same hometown (South Park).

(b) E_1: Kyle and Stan are were born in the same year.

E_2: Kyle and Stan are friends.

(c) E_1: A person grew up in Boston.

E_2: A person currently lives in Boston.

(d) E_1: A person was born 40 years ago.

E_2: A person is married.

7.7.12 [SN] It is common to compute some statistics about a small population of data and then use these to make predictions about a larger population. Assume that the basic information about the members in our example social network, provided in Table A.1, is representative of a much larger (expanded) social network. Use the data in that table to estimate the probabilities of the individual traits and then use the independence assumption and these individual probability estimates to calculate the probabilities requested below.

(a) What is the estimated probability of a random member in the larger social network being a female who lives in South Park?

(b) What is the estimated probability of a random member in the larger social network being a male born after 1990?

(c) What is the estimated probability of a random member in the larger social network being a female who currently lives in the hometown they grew up in?

7.7.13 [SN] In the previous exercise the requested joint probabilities were computed by calculating the probabilities of individual traits from the small population and then combining these using the independence assumption. In this exercise you should instead "directly" compute the requested joint probabilities from Table A.1. To make things clear the solution for the first part is provided.

(a) What is the estimated probability of a random member in the larger social network being a female who lives in South Park?

Solution: From Table A.1 we see that there are four people who live in South Park and none of the four are female. Since the table contains 27 people, our estimate of the requested probability that a person lives in South Park and is female is $0/27 = 0$. We would be more confident that this estimate would hold up for the larger population if the small population had more than 27 people and if we knew that the small population were representative of the larger population.

(b) What is the estimated probability of a random member in the larger social network being a male born after 1990?

(c) What is the estimated probability of a random member in the larger social network being a female who currently lives in the hometown they grew up in?

(d) Using this "direct" method, is the independence assumption needed and is it used to generate the solution? Why might this "direct" method not work as well in practice as the indirect method where we calculate only the individual probabilities from the data?

7.7.14 There are ten Democrats and five Republicans available to serve on a small two-person committee. If the committee is formed by randomly selecting two people from this pool of 15 people, then find the probability that:

(a) there is at least one Democrat on the committee.

(b) there is at least one Republican on the committee.

(c) there are only Democrats on the committee.

(d) there are only Republicans on the committee.

(e) the committee is made up of people from both parties.

7.7.15 In a box of light bulbs there are 12 good ones and 4 defective ones. If five bulbs are to be removed, find the probability that:

(a) All five are good.

(b) At most one bulb is bad.

7.7.16 A fair six-sided die is cast, a fair coin is tossed, and a card is drawn from a standard deck.

(a) What is the probability that you tossed a heads or drew an ace?

(b) What is the probability that you rolled a 4 or drew an ace?

(c) What is the probability that you rolled a 4, tossed a heads, and drew an ace?

Hint: First decide if these are disjoint or independent events.

7.7.17 You are given several six-sided and eight-sided dice. You are told to roll one six-sided die and if the result is a 1, 2, or 3, to then roll an eight-sided die but if the result is a 4, 5, or 6 to then roll a six-sided die. What is the probability that you roll snake-eyes (i.e., two 1's)?

7.7.18 What is the probability that in seven rolls of a six-sided die, a 1 appears at least five times?

7.7.19 There is a blackout and you have no flashlight, but need to put on some socks. You feel around in your sock drawer and find that there are six socks. If you only own blue and black socks, and if the probability that you pull out a pair of black socks is $\frac{2}{3}$, then what is the probability that you pull out a pair of blue socks? Justify your answer.

7.7.20 For each of the following questions, you are given a standard deck of cards, which is shuffled before the start of the question. Cards are not replaced into the deck during a question (i.e., if you pick two cards the first is not replaced before the second is selected).

(a) If I draw two cards, what is the probability that the both are aces?

(b) If I draw five cards, what is the probability of getting a flush? (Recall that a flush means that all cards are of the same suit.)

(c) If I draw a five-card poker hand, what is the probability that the hand contains no hearts?

7.7.21 If you draw a poker hand of five cards from a freshly mixed deck, what is the probability of the hand having more red cards than black cards? Show any work but also simplify your answer to get a numerical result.

7.7.22 You have one standard six-sided die. You roll it five times and record the results.

(a) How many possible outcomes are there?

(b) What is the probability that you get all 6's?

(c) What is the probability that you get exactly three 6's?

(d) What is the probability that none of the five rolls yields a 5 or 6?

(e) What is the probability that you get an even number on the first roll or on the second roll?

(f) If I add up the values of the five rolls, what value would I expect to get? That is, what is the expected value of the sum of these five rolls?

(g) Each of the five rolls is an event. Are the events independent? Are they disjoint?

7.7.23 What is the probability of flipping a coin ten times and getting a total of five heads and five tails?

7.7.24 A probability distribution shows the probabilities associated with each possible outcome. What is the probability distribution associated with rolling two dice and recording the sum of the two rolls?

7.7.25 A coin is flipped three times and the number of heads is recorded.

(a) Compute the probability distribution associated with this experiment and enter your results into a table. Convert the results to a decimal value.

(b) Manually run this experiment 30 times. That is, flip a coin three times, record the number of heads observed, and then repeat this procedure 30 times. Then calculate the observed probability for each case (e.g., if 2 heads came up 10 of the 30 times then the associated probability is 0.33) and enter it into the same table as for the previous part.

(c) Compare the "theoretical" results from part (a) with the "actual" results from part (b). Comment on any differences.

(d) Do you think that the differences would diminish, worsen, or stay the same if you repeated the experiment 100 times instead of 30 times?

7.7.26 (∗) You are a prisoner sentenced to death. However, you are given a chance to live by playing a simple game. You are given 50 black marbles, 50 white marbles and 2 empty bowls. You can divide these 100 marbles into these 2 bowls in any manner as long as you use all of the marbles. You will then be blindfolded and the bowls will be repeatedly swapped around and the marbles within each bowl will be mixed. You then will then be required to choose one bowl and remove one marble. If the marble is white you will live, but if the marble is black you will die. How do you divide up the 100 marbles between the two bowls to maximize your chance of living?

7.7.27 ($*$) You are on a game show. You are shown three doors, which have prizes behind them. One door has a great prize, such as a new car, while the other two doors have the same terrible prize, such as a goat. You select a door, but do not open it. The game show host then will open one of the doors that you did not choose that he knows to have the terrible prize. You then are given the choice to switch your choice to the other unopened door. Do you switch your door? That is, if you switch doors will your chance of winning the great prize go up, go down, or stay the same?

Chapter 8

Algorithms

> An algorithm must be seen to be believed.
>
> Donald Knuth
> *The Art of Computer Programming*
> *Volume I: Fundamental Algorithms*

An *algorithm* is a step-by-step procedure for carrying out a task. A recipe can be considered to be a simple algorithm, since it contains step-by-step instructions for the task of baking a cake. Of course, in this book we are primarily concerned with algorithms that can solve mathematical or computational problems and can be implemented on a modern digital computer. It is worth noting that algorithms can be used to *implement* the operations on many of the mathematical structures that we study in this book (e.g., set union), as well as various mathematical processes, such as logical inference. Algorithms can also be used to solve important, real-world problems. For example, graphs can be used to represent social networks and algorithms that operate on these graphs can determine useful things, such as the shortest connection between two people in Facebook and LinkedIn, via a chain of friends or colleagues. This chapter will introduce algorithms and provide several examples, but additional examples will be provided when we discuss graph algorithms in Chapter 8.

8.1 What is an Algorithm?

The concept of "algorithm" has been defined in several different ways. Each tends to emphasize a different aspect, or characteristic, of algorithms.

An algorithm is:

- A step-by-step procedure for carrying out a computation.
- An unambiguous computational procedure that takes some input and produces some output.
- A set of well-defined instructions for completing a task with a finite amount of effort in a finite amount of time.
- A set of instructions that can be mechanically performed in order to solve a problem.

These definitions make several important points. First, the procedure to be performed or executed must be unambiguous and *precisely* defined. So how should we specify our algorithms?

- One might consider using a natural language, such as English or French, to specify algorithms. However, the ambiguity inherent in natural languages means that such a specification will need to be lengthy, and not all that easy to understand. For example, think how difficult it would be to describe the quadratic formula in English, without using any algebraic notation whatsoever.
- The other extreme would be to use a computer language, such as C++ or Java. Since this text assumes no prior knowledge of computer programming, we can't take this approach. (Besides, which language should we use?)
- The happy medium is to use *pseudocode*, which is similar to a computer programming language (in terms of its control structures and precision), but is easily readable by non-programmers.[1]

Another key aspect of an algorithm is that it transforms a given input into some output; in this regard, algorithms are like the functions that we studied in Chapter 5. Finally, an algorithm should require only a finite number of steps to complete. This is actually a meaningful requirement, since there are many important computational problems whose "solution" never terminates for some inputs.

[1] Even computer professionals tend to use pseudocode when designing algorithms. Once the algorithm is finalized, it is usually fairly easy to translate the pseudocode into an actual computer program, written in a computer programming language.

8.2 Applications of Algorithms

Algorithms have many uses and applications. For one thing, they describe *how* to implement many of the operations that we study in this book, such as set membership, union, and intersection (Section 8.3.1 introduces two algorithms that implement the set membership query). This topic is so important that most computer science curricula dedicate an entire course to the study of *data structures* and the algorithms that operate on them. Without clever and efficient algorithms for operating on these structures, we would not be able to do many of the things that we take for granted, such as searching the Web or playing a video game.

Many other mathematical problems have algorithmic solutions. For example, there are algorithms to solve problems in propositional logic (see Section 3.1), and to compute most of the functions mentioned in Chapter 5. Of more practical importance is the fact that algorithms are routinely used to solve complex real-world problems. The following list of algorithms gives a mere glimpse at their diversity and ubiquity.

- The effectiveness of Google's search engine is largely due to its *PageRank* algorithm, which assigns a value to each web page based on its importance.

- *Prim's algorithm*, a graph-based algorithm described in Section 9.4.2, can be used by a cable company to determine how to connect all of the homes in a town using the least amount of cable.

- *Dijkstra's algorithm*, which finds the shortest path between all vertices in a graph, can be used to find the shortest route, by miles or travel cost, between a city and all other cities, or can find the shortest connection between all pairs of people in a social network.

- The *alpha-beta pruning algorithm* improves the performance of game playing (e.g., chess) programs by quickly eliminating moves that are provably sub-optimal.

- The *Sutherland-Hodgman polygon clipping algorithm* speeds up the rendering of images for computer graphics and video game programs by removing objects that do not fall into the "camera's" field of view.

- The RSA *encryption algorithm* implements public-key cryptography, and, along with other similar algorithms, makes e-commerce possible by allowing for secure transactions.

Algorithms have been used for thousands of years; in principle, they can be implemented by a person using pencil and paper. However, work on algorithms exploded with the development of modern digital computers. The tremendous speed of these machines made it easy to develop algorithms that could be applied to significant problems.

However, it is worth noting that there are large classes of real-world problems for which the only known algorithms cannot complete, even on today's fast computers, in a reasonable amount of time. Perhaps more surprisingly, it has been *proven* that there are problems for which there is no algorithmic solution whatsoever!

Remark. An algorithm that would require infinitely many steps to give us its answer would be pretty useless. So, it would be nice if somebody could develop an algorithm that could take an arbitrary computer program as its input, producing an output "yes" if said program terminates after finitely many steps, and "no" otherwise (i.e., if said program will never terminate). The problem of finding such an algorithm is called the *halting problem.* In 1936, Alan Turing[2] proved that the halting problem is *undecidable*, meaning that no algorithm, no matter how clever and sophisticated, can answer this question. □

8.3 Searching and Sorting Algorithms

In this section we describe and analyze important algorithms for searching and for sorting. These algorithms, which are amongst the most heavily-studied classes of algorithms in computer science, serve as excellent examples of algorithms.

8.3.1 Search Algorithms

Search algorithms are an important class of algorithms in computer science because of the need to quickly retrieve information. In this section we describe two algorithms: linear search and binary search. Linear search is very straightforward but inefficient, whereas binary search is slightly more complex, but much more efficient. Both of these algorithms allow us to search for a specific element in a set or list and thus both are suitable for

[2] Turing, A. M., "On computable numbers, with an application to the Entscheidungsproblem," *Proceedings of the London Mathematical Society*, Series 2, 42 (1936), pp 230–265.

handling the set membership query. The search problem is to determine if an element x occurs in the list L or set S.

In the remainder of this section we assume that the search operates on a list, rather than a set. The difference between the two structures is that in a set the elements are unordered, whereas in a list the elements have an order. Thus, in a list, there is such a thing as a first element, second element, etc., whereas no such concept exists for sets. However, since sets can be implemented as lists (we just ignore the ordering information), the algorithms in this section apply equally to both structures.

8.3.1.1 Linear Search

A very simple algorithm for determining whether element x occurs in any of the n elements in the list L is provided below. In the algorithm below, $L[i]$ refers to the ith element in the list.

The Linear Search Algorithm

1. **repeat** as i varies from 1 to n
2.   If $L[i] = x$ then return "FOUND" and stop
3. return "NOT FOUND"

This algorithm checks each element in L, starting with the first one and continuing until the nth one is checked. That is, the **repeat** statement in step 1 causes step 2 to repeat up to n times, where the value i increases by 1 each time. If a match is ever found, the algorithm immediately returns "FOUND" and terminates. If the algorithm checks all elements and no match is found, then it returns "NOT FOUND."

This algorithm potentially requires each of the n elements in the list to be checked, which means that it would not be practical for repeatedly searching very long lists, such as a phone book. In that case, the more efficient binary search would be preferred, which we describe in the next section.

Example 8.1: [SN] How many comparisons will the linear search algorithm perform before it finds Niko in Table A.1 on page 318, which provides the basic information about members in our example social network?

Solution: The table will be searched in order starting from the beginning and since Niko is the twentieth member in the table, 20 comparisons will be needed. □

Example 8.2: [SN] In 2011, Facebook claimed that it had 750 million active users. Assuming that this number is accurate and that the member list is stored alphabetically by last name and then first name, about how long would linear search take to find Zysel Zywiec if the computer can perform one comparison in one nanosecond (i.e., one-billionth or 10^{-9} seconds)?

Solution: We can safely assume that Zysel Zywiec is very close to the last entry in the alphabetized member list. Thus, the linear search for Zysel will take:

$$750{,}000{,}000 \times 10^{-9} \text{ sec} = 0.75 \text{ sec}$$

This example demonstrates that today's fast computers can search long lists, even with inefficient algorithms. But if many searches are required, even 0.75 seconds may be too slow. As is demonstrated later in Example 8.10, Zysel can be found using binary search in only 29 nanoseconds. □

8.3.1.2 Binary Search

The binary search algorithm enables us to search for an element x in a *sorted* n-element list L without checking each element. This is only possible because the elements in L are sorted; had they not been sorted, this would not be possible. The binary search algorithm is described below.

The Binary Search Algorithm

1. Initialize $min \leftarrow 1$ and $max \leftarrow n$
2. **Repeat** until $min > max$
3. $\quad midpoint = \frac{1}{2}(min + max)$
4. $\quad$ Compare x to $L[midpoint]$
   - (a) if $x = L[midpoint]$ then return "FOUND"
   - (b) if $x > L[midpoint]$ then $min \leftarrow midpoint + 1$
   - (c) if $x < L[midpoint]$ then $max \leftarrow midpoint - 1$
5. return "NOT FOUND"

Binary search operates as follows. The algorithm maintains the values for *min* and *max*, so that they always specify the "window" of values that may contain x. Thus this window initially contains all of the elements in L (step 1). The algorithm operates by repeatedly cutting this window in half. This is controlled by the **repeat** statement (step 2), which causes steps 3 and 4 to repeat until $min > max$, indicating that there are no elements left in the window. When this happens the algorithm returns "NOT FOUND" (step 5). If the window does have elements in it, then the algorithm calculates the midpoint of the window (step 3) and then compares the value of x to the value at this midpoint (step 4). If the midpoint value equals x, then the algorithm immediately returns "FOUND" and terminates (step 4a). Otherwise the value at the midpoint is used to determine whether x occurs in the half of the list to the right or left of the midpoint. If x is greater than the midpoint value, then x, if it occurs at all, must occur in the right half, so we set *min* accordingly (step 4b); otherwise, if x is less than the midpoint value, then x must occur in the left half of the list, and so *max* is set accordingly (step 4c).

Example 8.3: Use binary search to find the element 4 in the sorted list (1 3 4 5 6 7 8 9). For each iteration of the algorithm show the values of *min*, *max*, and *midpoint* after step 3 executes. Also, how many values are ultimately compared to "4"?

Solution:

1. Initially $min = 1$ and $max = 8$. The midpoint value is calculated as $\frac{1}{2}(1 + 8) = 4$ (we always round down). The value at the midpoint is 5; since $4 < 5$ we execute step 4c and set *max* to $midpoint - 1 = 3$.

2. Now $min = 1$ and $max = 3$. The midpoint value is calculated as $\frac{1}{2}(1 + 3) = 2$. The value at the midpoint is 3; since $4 > 3$ we execute step 4b and set *min* to $midpoint + 1 = 3$.

3. Now $min = 3$ and $max = 3$. The midpoint value is calculated as $\frac{1}{2}(3 + 3) = 3$. The value at the *midpoint* is 4. Since $4 = 4$ we execute step 4 and return "FOUND."

During the execution of the algorithm we only check three values: 3, 4, and 5. Because we cut the list in half each iteration of the algorithm, the list will shrink quite quickly and hence the search will complete quickly. In Section 8.4.3.2, we discuss how many comparisons are required in terms of the input length n of the list. □

Example 8.4: Repeat Example 8.3 but this time search for the element 2 in the ordered list (1 3 4 5 6 7 8 9).

Solution:

1. Initially $min = 1$ and $max = 8$. The midpoint value is calculated as $\frac{1}{2}(1+8) = 4$ (we always round down). The value at the midpoint is 5; since $2 < 5$ we execute step 4c and set *max* to $midpoint - 1 = 3$.

2. Now $min = 1$ and $max = 3$. The midpoint value is calculated as $\frac{1}{2}(1+3) = 2$. The value at the midpoint is 3; since $2 < 3$ we execute step 4c and set *max* to $midpoint - 1 = 1$.

3. Now $min = 1$ and $max = 1$. The midpoint value is calculated as $\frac{1}{2}(1+1) = 1$. The value at the *midpoint* is 1. Since $2 > 1$ we execute step 4b and set *min* to $midpoint + 1 = 2$.

4. Now $min = 2$ and $max = 1$. Since $min > max$, the "Repeat until" statement on line 2 terminates and we continue processing on line 5. Thus we return "NOT FOUND."

For this example only the values 5, 3, and 1 are checked. □

8.3.2 Sorting Algorithms

Sorting is another heavily-studied topic in computer science; significant effort has gone into developing effective sorting algorithms. Sorting is important because many tasks, particularly searching, are much easier to do with sorted lists than with unsorted lists. After all, think about how difficult it would be to locate someone in an unsorted phone book! In this section we will use the simple example below to show how algorithms can solve the sorting problem.

Example 8.5: Sort the list of numbers (9 2 8 4 1 3) into increasing order.

Solution: There are many algorithms for solving this problem, but all will yield the same solution: (1 2 3 4 8 9). □

In the remainder of this section we will look at two sorting algorithms. Although we assume that the goal is to sort the lists into ascending order, it would be trivial to modify the algorithms to handle descending order or alphabetical order.

8.3.2.1 Bubblesort

Bubblesort works by repeatedly scanning through a list, each time allowing the largest element in the unsorted part of the list to "bubble" to the end of that part of the list. Thus, after one iteration, the largest element will be at the end of the list and after two iterations the second largest element will be the second-to-the-last element in the list (and the largest element will still be at the end of the list). Note that after each iteration of the bubblesort algorithm we need to process one fewer element on the list, since after k iterations the last k items on the list will already be sorted. The "bubbling up" process is implemented by repeatedly comparing consecutive elements in the list and swapping the order of those two elements if they are not already in the desired order. The bubblesort algorithm, which sorts the n-element list $L = (l_1, l_2, \ldots, l_n)$, is summarized below. We immediately follow the algorithm with an example that shows it in action.

The Bubblesort Algorithm

1. **repeat** as i varies from n down to 2
2.   **repeat** as j varies from 1 to $i - 1$
3.   swap l_j with l_{j+1} if $l_j > l_{j+1}$

The **repeat** statement that spans steps 2 and 3 does most of the work. It compares pairs of consecutive elements and swaps them if the values are out of order (i.e., if the first element is larger than the second). Each time that repeat statement is hit, it compares l_1 with l_2, then l_2 with l_3, then l_3 with l_4, up to l_{i-1} with l_i. The first **repeat** statement at step 1 determines the last element to check in the list and is responsible for making sure that one less element is checked each time the **repeat** statement at step 2 is executed.

Example 8.6: Use bubblesort to sort the list of numbers (9 2 8 4 1 3) into increasing order. Show the intermediate results after each step.

Solution: In the solution below, the two values being compared at each step are highlighted in bold and any elements that have been completely sorted are underlined. The underlined values never need to be checked again.

Iteration 1:
$(\mathbf{9}\ \mathbf{2}\ 8\ 4\ 1\ 3) \to (\mathbf{2}\ \mathbf{9}\ 8\ 4\ 1\ 3)$, swap since $9 > 2$
$(2\ \mathbf{9}\ \mathbf{8}\ 4\ 1\ 3) \to (2\ \mathbf{8}\ \mathbf{9}\ 4\ 1\ 3)$, swap since $9 > 8$
$(2\ 8\ \mathbf{9}\ \mathbf{4}\ 1\ 3) \to (2\ 8\ \mathbf{4}\ \mathbf{9}\ 1\ 3)$, swap since $9 > 4$
$(2\ 8\ 4\ \mathbf{9}\ \mathbf{1}\ 3) \to (2\ 8\ 4\ \mathbf{1}\ \mathbf{9}\ 3)$, swap since $9 > 1$
$(2\ 8\ 4\ 1\ \mathbf{9}\ \mathbf{3}) \to (2\ 8\ 4\ 1\ \mathbf{3}\ \underline{\mathbf{9}})$, swap since $9 > 3$

Iteration 2:
$(\mathbf{2}\ \mathbf{8}\ 4\ 1\ 3\ \underline{9}) \to (\mathbf{2}\ \mathbf{8}\ 4\ 1\ 3\ \underline{9})$, don't swap since $2 < 8$
$(2\ \mathbf{8}\ \mathbf{4}\ 1\ 3\ \underline{9}) \to (2\ \mathbf{4}\ \mathbf{8}\ 1\ 3\ \underline{9})$, swap since $8 > 4$
$(2\ 4\ \mathbf{8}\ \mathbf{1}\ 3\ \underline{9}) \to (2\ 4\ \mathbf{1}\ \mathbf{8}\ 3\ \underline{9})$, swap since $8 > 1$
$(2\ 4\ 1\ \mathbf{8}\ \mathbf{3}\ \underline{9}) \to (2\ 4\ 1\ \mathbf{3}\ \underline{\mathbf{8}}\ \underline{9})$, swap since $8 > 3$

Iteration 3:
$(\mathbf{2}\ \mathbf{4}\ 1\ 3\ \underline{8\ 9}) \to (\mathbf{2}\ \mathbf{4}\ 1\ 3\ \underline{8\ 9})$
$(2\ \mathbf{4}\ \mathbf{1}\ 3\ \underline{8\ 9}) \to (2\ \mathbf{1}\ \mathbf{4}\ 3\ \underline{8\ 9})$
$(2\ 1\ \mathbf{4}\ \mathbf{3}\ \underline{8\ 9}) \to (2\ 1\ \mathbf{3}\ \underline{\mathbf{4}}\ \underline{8\ 9})$

Iteration 4:
$(\mathbf{2}\ \mathbf{1}\ 3\ \underline{4\ 8\ 9}) \to (\mathbf{1}\ \mathbf{2}\ 3\ \underline{4\ 8\ 9})$
$(1\ \mathbf{2}\ \mathbf{3}\ \underline{4\ 8\ 9}) \to (1\ \mathbf{2}\ \underline{\mathbf{3}}\ \underline{4\ 8\ 9})$

Iteration 5:
$(\mathbf{1}\ \mathbf{2}\ \underline{3\ 4\ 8\ 9}) \to (\mathbf{1}\ \underline{\mathbf{2}}\ \underline{3\ 4\ 8\ 9})$ □

This example should make it very clear how bubblesort works. Note that in the **repeat** loop, the "counting variable" i takes all the values in the set $\{2, 3, \ldots, n - 1\}$, albeit in backwards order; this is a set of size $n - 1$. In other words, there are $n - 1$ iterations in the **repeat** loop. You might wonder why aren't there n iterations. In other words, why doesn't i take the values in the set $\{1, 2, \ldots, n\}$? The reason for this is based on the fact that each iteration of the **repeat** loop puts an additional element into its proper place in the final sorted list. After $n - 1$ iterations, the last $n - 1$ elements are in their proper place. So there's only one element left to place into its proper place, and only one place that hasn't been filled; there's no other slot for this unclaimed element fill, except for this as-yet unfilled slot.

You might wonder how much work bubblesort requires to sort a set of size n. The answer is complicated by the fact that we need to be more precise in determining exactly what it is that we're trying to measure. It's also (somewhat) complicated by the fact that at each successive iteration, there's less work to do (since some of the list is already sorted). In Section 8.4.3.3 we will come back to this issue and compute exactly how many comparisons bubblesort must do in terms of n, the number of elements in the original unsorted list.

8.3.2.2 Mergesort

Mergesort sorts a list in a very different manner from bubblesort. It uses a *divide-and-conquer* approach, which means that it divides the sorting problem into smaller sorting problems, solves the smaller problems, and then combines these solutions to form a solution to the original problem. While quite simple, this algorithm is also ingenious, because, as we shall see in Section 8.4.3.3, it works far more quickly than bubblesort. The mergesort algorithm for sorting a list L is described below. In this case we describe it as a function.

The Mergesort Algorithm

function mergesort(L)

1. If L has one element then return L; otherwise continue.
2. $l_1 \leftarrow$ mergesort(left half of L)
3. $l_2 \leftarrow$ mergesort(right half of L)
4. $L \leftarrow$ merge(l_1, l_2)
5. return(L)

Mergesort will always return a sorted list. If the list passed into mergesort contains only one element (step 1), then the list is trivially sorted so mergesort just returns that list. Otherwise mergesort calls *itself* to sort the left half of the list (step 2) and then the right half of the list (step 3). It then takes these two sorted sublists and merges them into a single composite sorted list (step 4) and then returns the resulting list (step 5).

In computer science lingo, mergesort is a *recursive* function since it invokes itself. At first glance, this may seem improper—like defining a word in terms of itself—but this is not a problem. After all, the recursive calls are on successively smaller lists, which means that eventually the list will contain a single element and the sequence of recursive calls will terminate. This is analogous to the recursive sequences we studied in Chapter 2, in which a term at a given point in the sequence is defined in terms of previous terms in the sequence.

Before we apply mergesort to an example, we must address two other points. First, if the list to be sorted contains an odd number of elements, then the list cannot be split evenly in halves so we will keep the "extra" element with the left half of the list. Second, the task of merging two sorted lists (in step 4) calls another function, *merge*, which implements the merge algorithm. We guarantee that you have used this algorithm before. To merge two sorted lists you place your finger on the first element of each list and then determine which of the two elements belongs first. Once that is determined you copy that element to the sorted list, advance your finger on the original list to the next element, and then repeat the steps until all elements have been processed.

We can now move on to an example that should clarify the mergesort algorithm. We will use the same example as we used for the bubblesort algorithm.

Example 8.7: Provide a step-by-step description of how mergesort would sort the list of numbers (9 2 8 4 1 3) into increasing order.

Solution: A step-by-step visual depiction of how mergesort works is provided below. To help show what is going on we highlight the list elements in bold after they have been sorted and are about to be merged together (step 4). The first two lines below sort the left hand side of the original list, and the second set of two lines sorts the right hand side of the original list and merges the two sublists together.

[9 2 8 4 1 3] → [9 2 8] [4 1 3] → [9 2] [8] [4 1 3]

[**9**] [**2**] [8] [4 1 3] → [**2 9**] [**8**] [4 1 3] → [2 8 9] [4 1 3]

[2 8 9] [4 1 3] → [2 8 9] [4 1] [3] → [2 8 9] [**4**] [**1**] [3]

[2 8 9] [**1 4**] [**3**] → [**2 8 9**] [**1 3 4**] → [1 2 3 4 8 9] □

To help clarify further what is going on, we briefly describe the operations corresponding to the first two lines in the visual depiction. On the first line, the original list is split in two and then the first half is split again. On the second line the leftmost list is split again, so that each list has only a single element, which means that we consider each of these one-element lists to be already sorted. These two lists are then merged, which in this case just means swapping the single elements that comprise each list (i.e., the 9 and 2 are swapped). At this point the list next to it, which contains an

"8", is sorted since it has only one element, so these two lists are merged, yielding a list with (2 8 9). The remainder of the list is sorted in a similar manner. □

8.4 Analysis of Algorithms

An algorithm is a set of instructions that solves a particular problem for all legitimate input instances. However, not all algorithms that solve a class of problems are equally good; some algorithms are more efficient than others. The task of determining the efficiency of algorithms is referred to as the *analysis of algorithms*. This is a vast topic of study and ongoing research; we can only give a basic overview in this book. However this section will introduce you to a few techniques that will allow you to analyze simple algorithms.

8.4.1 How Do We Measure Efficiency?

Algorithms are described using an unambiguous set of instructions. As mentioned above, these instructions may be written in a computer language, in pseudocode, or in a natural language such as English. A human can perform all of the steps of an algorithm, as we saw when working through the searching and sorting examples earlier in this chapter. As humans, what do we care about with respect to the performance of an algorithm? The main consideration for most of us is *time*—which is why none of the homework questions at the end of this chapter asks you to sort a long list of numbers! Traditionally, most work on algorithmic efficiency relates to the time it takes for the algorithm to complete and hence that is what we focus on in this section.

It is worth briefly noting, however, that the other resource that is sometimes analyzed is the *space* required by an algorithm. When a human manually executes an algorithm it might be reasonable to measure the total amount of paper needed for the calculations. But in computer science the main measure related to space is the maximum amount of computer memory that is needed at any one time during the execution of the algorithm. Although we are only going to consider the time it takes to execute an algorithm, it is important to note that space is the more limiting resource for some classes of problems.

8.4.2 The Time Complexity of an Algorithm

There are many possible ways to measure the time required by an algorithm to solve a problem. The most direct way is just to run the algorithm on a computer and measure the time it takes for the algorithm to complete. However, there are several problems with this approach. First, there are many types of computers, each with different speeds, so which one should we use? One possible solution is to pick a single benchmark computer system and use that to evaluate all algorithms, but this is not very practical, as computer technology changes over time. Beyond this, some computers are optimized to perform certain operations quickly, so the choice of a computer architecture would bias the performance results. Even worse, the specific input that we choose to measure performance on might impact the execution time of the algorithm, so how do we choose the input on which to evaluate our performance? For these and other reasons, the analysis of algorithms does not rely on actual computer execution time to measure performance. Instead, computer scientists focus on the *time complexity* (sometimes called the *runtime complexity*) of an algorithm.

The run-time complexity of an algorithm describes how the number of steps, or operations, *grows* as the length of the input grows. If we implemented bubblesort and mergesort as functions in a programming language, we could execute them on a computer on input lists of various lengths and record the average number of operations required to sort the lists. The results would look something like the results in Table 8.1, where bubblesortOps(n) and mergesortOps(n) are the functions that describe how the number of operations carried out by their respective algorithms varies with the input length of the list.

n	2	4	8	16	32	64
bubblesortOps(n)	4	16	64	256	1024	4096
mergesortOps(n)	2	8	24	64	160	384

Table 8.1: Growth Function for Bubblesort and Mergesort

Note that the running time of the algorithm can be viewed as a function of the length of the input to the algorithm and hence evaluating the time complexity of an algorithm relates to the *growth of functions*, which we studied in Section 5.7.2. The data in Table 8.1 specifies the number of operations for input lists with widely-varying lengths n, so that we get a good idea of how these functions, which measure the number of operations, grow with respect to n.

Based on the data in Table 8.1, it seems clear that mergesort is be the better algorithm, since it requires fewer operations to sort a list of a given input length. This is even clearer if we look at the data graphically in Figure 8.1. From the figure it should be clear that the differences between the two algorithms are accelerating and that if the input list kept growing, these differences would become even larger.

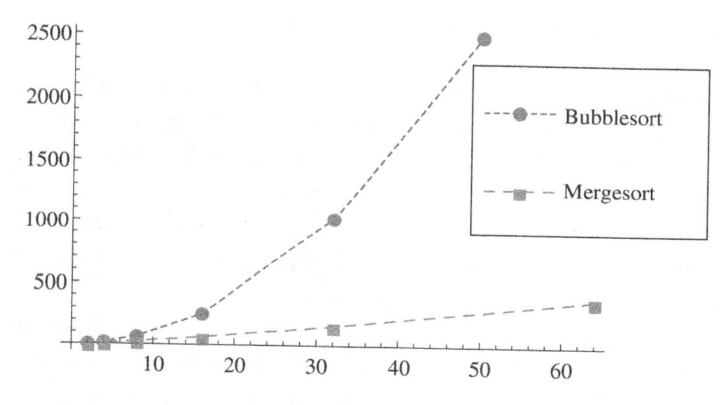

Figure 8.1: Running Time of Bubblesort and Mergesort

If we look at the data long enough and carefully enough, we might find a closed formula that relates n and the running times of the two algorithms. This is exactly the type of problem that we addressed in Chapter 2, where we found closed formulas for various sequences. The following problem asks you to find the closed formulas. It should not be too difficult to find the closed formula for bubblesort, although the closed formula for mergesort may prove to be more difficult.

Example 8.8: Using the data in Table 8.1, find closed formulas for the functions bubblesortOps(n) and mergesortOps(n), or equivalently, for the sequences associated with these two functions.

Solution: Looking at the bubblesort data in the table, we can easily see that bubblesortOps$(n) = n^2$. The closed formula for mergesortOps(n) is not nearly as obvious. As it turns out, the closed formula is $a_n = n \log_2 n$, where "$\log_2$" denotes the base 2 *logarithm* of a number. Simply put, $\log_2 n$ equals the value x such that $2^x = n$. Thus $\log_2 2 = 1$ since $2^1 = 2$ and $\log_2 256 = 8$ since $2^8 = 256$. If you plug in the values of n into the formula $a_n = n \log_2 n$, you will see that we generate the sequence that matches the data for mergesort in Table 8.1. □

It turns out that the data in Table 8.1 is representative of how the bubblesort and mergesort algorithms behave in general:

- bubblesortOps(n) grows like n^2 as n gets larger and larger. We will show why this is so in Section 8.4.3.3.

- mergesortOps(n) grows as $n \log_2 n$. We do not provide a detailed derivation of this result, as it is somewhat complicated; however, we shall briefly discuss the analysis in Section 8.4.3.4.

So the running time of bubblesort is quadratic in the length n of the input, which is a fancy way of saying that the running time grows proportionally to n^2. Moreover, the running time of mergesort is proportional to $n \log_2 n$. The running time of an $n \log_2 n$ sorting algorithm (such as mergesort) grows *much* more slowly than that of an n^2 algorithm (such as bubblesort); once again, see Figure 8.1, which shows how these running times diverge even for relatively small values of n.

Thus for large lists (i.e., large values of n), mergesort is much quicker than bubblesort. Unless we have a guarantee that we'll never have to sort a large list, we should never use a sorting algorithm with quadratic running time (such as bubblesort).

We hope that the reader appreciates the significance of the divide-and-conquer approach. The fact that mergesort performs better than bubblesort relies on the fact that it is less work to recursively sort (and then merge) two lists of size $n/2$ than it is to use bubblesort on a single list of size n. This is a very significant insight, which is quite counterintuitive for most people.

8.4.3 Analysis of Several Algorithms

We are now ready to discuss the analysis of the algorithms that we studied for searching and sorting. As a slight simplification, rather than looking at the total number of operations that the algorithm needs, we will only count the number of comparisons. Although these algorithms use other operations (for instance, data gets moved around when sorting), the number of comparisons turns out to be the dominant term of the overall complexity.

Before going on, we should point out that we are focusing on the *worst-case performance* of the algorithm. That is, the linear search algorithm takes n comparisons only when the element we are searching for is not in the list or is the last element in the list; if the desired element is earlier in the list, then linear search will take less time. Since this worst-case setting provides the strongest assurance to the user (an iron-clad guarantee of how well the algorithm will perform, if you will), computer scientists tend to concentrate on worst-case performance when analyzing algorithms. However, algorithms can be analyzed under other settings as well. We often want to know the *average-case performance* of an algorithm; occasionally, we even want to know the *best-case performance*.

8.4.3.1 Analysis of the Linear Search Algorithm

The linear search algorithm is very simple, iterating through the elements of the list, checking to see whether one of them matches the data item that we seek. In the worst case, we would have to check every element of the list (this happens when the data item we seek is either the final element in the list or when it's not in the list at all). If there are n elements in the list, this means that the algorithm requires n comparisons. This means that if we let $\text{linearseachOps}(n)$ denote the worst case number of comparisons that linear search uses to search a list of size n, then

$$\text{linearseachOps}(n) = n.$$

Note that the number of comparisons is linear in the input length of the list.

Example 8.9: If you use linear search to search for a specific name in a list of names, what is the best-case performance and the average-case performance of the algorithm?

Solution: The best-case performance occurs when the name you are searching for is the first name on the list, in which case exactly one comparison is used. The average-case performance cannot be calculated unless you make some assumptions about the name being searched on. If we assume that the name must be on the list and that it is as likely to be in any position as any other, then, on average, one would expect the algorithm to require about $n/2$ comparisons. □

8.4.3.2 Analysis of the Binary Search Algorithm

The binary search algorithm, as described in Section 8.3.1.2, repeatedly cuts the list to be searched in half. Since only one comparison is required each time the list is cut in half, the number of comparisons needed will equal the number of iterations required. In the worst case, the algorithm will need to iterate (and cut the list in half) until it contains only a single element. This will require $\log_2 n$ iterations, where, as before, $\log_2 n$ is the value x such that $2^x = n$.

In case this answer is not obvious to you, we will motivate it using an example. Suppose the list to be searched contains 8 elements, like the list in Example 8.3. If we cut the list in half once, we get 4 elements. If we cut it in half a second time, then we get 2 elements, and finally, if we cut it in half a third time, we get 1 element. Thus, it takes 3 cuts to go from 8 elements to 1. Since $\log_2 8 = 3$ (i.e., $2^3 = 8$), we see that the formula works.

Remark. The actual number of comparisons is $\lceil \log_2 (n+1) \rceil$, where the ceiling function $\lceil \cdot \rceil$ may be found on page 184 of Section 5.7.1. □

Let binaryseachOps(n) denote the number of comparisons that binary search requires when searching a sorted list of n elements. We have shown that

$$\text{binaryseachOps}(n) \text{ is roughly } \log_2 n.$$

Since the linear search algorithm requires n comparisons versus roughly $\log_2 n$ for the binary search algorithm, the binary search algorithm is much more efficient. However, note that there's a hidden "overhead" cost associated with binary search, namely, that the binary search algorithm does require that the list be sorted. Sorting can be done using $n \log_2 n$ operations using mergesort (as discussed on page 270 and in Section 8.4.3.4), so sorting a list and then using binary search just once would require about $n \log_2 n + \log_2 n$ operations, which is more operations than linear search, which requires n operations (since $n \log_2 n > n$). Thus it makes no sense to sort a list just so that you can search it using binary search, unless you plan to search it many times. But it is very common to search a list many times, and for this reason most search algorithms assume that the list will be sorted first.

Example 8.10: [SN] In Example 8.2 we showed that if a computer can perform one comparison in 1 nanosecond, then using linear search it would take about 0.75 seconds to find Zysel Zywiec in an alphabetical member list of a social network with 750 million members. About how long would it take using binary search?

Solution: Binary search would require about $\log_2 750{,}000{,}000 = 29.48$ comparisons to find Zysel. At 10^{-9} seconds, or 1 nanosecond, per comparison, that would require only 29 nanoseconds. Clearly this is a big improvement over the 0.75 seconds (750,000,000 nanoseconds) required by linear search. □

Remark. [SN] Binary search is fine for nearly-static data sets, i.e., for sets that don't change very much. But we often have find items inside rapidly-changing data sets. For example, the social network in the Appendix is not carved in stone; it's pretty likely to change as new friends are added (and dropped). Moreover, this little social network is only an example. There are Facebook users with thousands of friends, and LinkedIn users with thousands of contacts. Moreover, Facebook (or LinkedIn) users often look to see whether somebody they know from elsewhere (say, in meatspace[3]) is a member. These are rapidly changing mega-social networks, having hundreds of millions of members. We can neither afford to sort the membership list each time it changes, nor do a linear search every time we want to find somebody. For problems such as these, where the data set is rapidly changing but we need to do many searches, more sophisticated techniques are needed. Alas, such techniques are beyond the scope of this book. □

8.4.3.3 Analysis of the Bubblesort Algorithm

We want to determine how many comparisons bubblesort must do to sort a list. We can determine the number of comparisons by carefully analyzing the bubblesort example from Example 8.6 and generalizing from it. If we look back at the number of comparisons that were performed in Example 8.6, we see that to sort the 5 items in the list (9 2 8 4 1 3) we used $5+4+3+2+1 = 15$ comparisons. If we think a little bit, we can generalize this for values of n other than 5. It will also help when generalizing to look back at the description of the bubblesort algorithm from Section 8.3.2.1. Computer scientists generally determine running times by looking at the details and structure of an algorithm.

The bubblesort algorithm, as described in Section 8.3.2.1, has two steps. The first step involves a repeat statement that executes n times. Each of those n times, a series of comparisons is done. The first time there will be $n-1$ comparisons and each time after that the number of comparisons will decrease by 1. Thus we can see that *generally*, a list with n elements will require $(n-1)+n-2+\cdots+2+1$ comparisons. If we let bubblesortOps(n) denote the number of comparisons that bubblesort uses when sorting a list of n elements, we see that

$$\text{bubblesortOps}(n) = \sum_{i=1}^{n-1} i \qquad (8.1)$$

[3] As opposed to cyberspace.

However, we can calculate this summation using a closed formula, since we know from Section 2.4.2 that

$$\sum_{i=1}^{n} i = \tfrac{1}{2}n(n+1) \qquad (8.2)$$

Since the summation in 8.1 only goes up to $n-1$ and not n, we can substitute $n-1$ for n in equation (8.2) to find that the total number of comparisons required to sort n items using bubblesort is given by

$$\text{bubblesortOps}(n) = \tfrac{1}{2}n(n-1) = \tfrac{1}{2}(n^2 - n) \qquad (8.3)$$

Example 8.11: Compute the number of comparisons required by bubblesort to sort lists with the following number of elements: 10, 100, 1,000, 10,000, and 100,000. Furthermore, if we ignore all of the bubblesort work except for the comparisons, how long would it take to sort a list with 100,000 elements using bubblesort on your PC?

For this example assume that your PC's processor operates at a relatively slow 1 GHz, which means that it executes one billion instructions per second. Also assume that a comparison can be accomplished in the time it takes to execute one instruction.

Solution: To determine the number of comparisons, we simply need to plug the values into equation (8.3). The results are shown in the table below.

# elements	10	100	1,000	10,000	100,000
# comparisons	45	4,950	49,9500	49,995,000	4,999,950,000

Clearly the work required by bubblesort grows quite rapidly with n. If 1 billion (10^9) instructions can execute each second, then one instruction will take one-billionth (10^{-9}) of a second. Thus it would take about 5 seconds to sort a 100,000-element list on a conventional single processor computer (at the moment four-processor "quad core" computers are quite common). □

Remark. In the initial version of this example, one of the authors had each instruction taking 0.01 milliseconds (10^{-5} seconds). In that case it took 50,000 seconds, or about 13.9 hours, to sort the 100,000-element list. The intended lesson was that inefficient algorithms like bubblesort cannot handle problems such as these in reasonable time. But with the more accurate processor speed in the current version of this example, which has the processor operating 1,000 times faster, that lesson is ruined! In this case the particular author made up the original instruction cycle time without much thought. But when was that original processor speed reasonable? We know that processor speeds have been doubling every two years or so (see the discussion of "Moore's law" in Section 3.1.10 on page 105) and thus we can expect a 1,000-fold increase every 20 years. Thus, the original processor speed was reasonable when the author studied computer architecture in college (a bit more than 20 years ago). So, what is the significance of the original mistake? First, it makes that particular author feel really old. But it also shows that tasks that one *human* generation ago could not be accomplished via brute force methods can now be accomplished by such methods. But the initial lesson concerning algorithmic complexity still holds; if the list to be sorted contained 100,000,000 items then even today's faster processor would have taken approximately 13.9 hours to sort the list. □

8.4.3.4 Analysis of the Mergesort Algorithm (∗)

How could we determine the time complexity of mergesort? For simplicity's sake, let a_n denote the number of comparisons that mergesort needs to sort a set of size n. Looking back at page 265, we see that if $n = 0$, then no comparisons are done at all; thus $a_1 = 0$. So let's suppose that $n > 1$. To keep things simple, let's suppose that n is even. How many comparisons are required by steps 2, 3, and 4?

- Step 2 asks us to sort the left half of the original list. Since the original list has n elements, the left half has $n/2$ elements. Just as a_n is the number of comparisons needed to sort a list of size n, we see that $a_{n/2}$ is the number of comparisons needed to sort a list of size $n/2$. So the cost of Step 2 is $a_{n/2}$ comparisons.

- Step 3 asks us to sort the right half of the original list. As in the previous bullet item, this can be done in $a_{n/2}$ comparisons.

- Step 4 asks us to merge the two sorted lists. It is not too difficult to see that since there are $n/2 + n/2 = n$ elements overall in the two lists, we can merge the two lists using at most $n - 1$ comparisons.

Thus we see that

$$\begin{aligned} a_n &= 2a_{n/2} + n - 1 \qquad \text{if } n \text{ is even and } n \geq 2, \\ a_1 &= 0. \end{aligned}$$

This is a recurrence relation, but a good deal more complicated than the kinds of recurrence relations you studied in Chapter 2. Using mathematical induction (as in Section 2.4), it is possible to show that

$$a_n = n \log_2 n - n + 1 \qquad \text{if } n \text{ is a power of 2.}$$

The formula for the case where n is not a power of 2 is a good deal more complicated than this; however, it is not too difficult to extend the analysis given above to cover general values of n, to show that the runtime of mergesort is roughly $n \log_2 n$; in terms of the O-notation we will introduce in Section 8.4.4, we would say that the mergesort has runtime $O(n \log_2 n)$. In conclusion, if we let $\text{mergesortOps}(n)$ denote the number of comparisons that mergesort uses when sorting a list of n elements, we see that

$$\text{mergesortOps}(n) \text{ is roughly } n \log_2 n.$$

8.4.4 Big-O Notation (∗)

In the previous section we characterized the running time of an algorithm by the number of operations[4] the algorithm requires in the worst case. But as we discussed in Section 8.4.2, computer scientists typically do not focus on the the actual number of operations the algorithm uses. Why not?

- The number of operations may depend on extraneous factors, such
  - as the skill of the programmer who implements the algorithm, or
  - the quality of the compiler that translates the program's source code into a runnable program.
- It may be too difficult to get an exact formula for the number of operations.
- There are cases where the exact formula is so complicated that it obscures what's really going on.

Instead, computer scientists are more concerned with how the number of operations grows as the input to the algorithm grows. Hence the time complexity of an algorithm is typically characterized in terms of its *asymptotic upper bound*, using the O (or Big-O) notation described in this section. Before we provide a mathematically rigorous definition, we will give some

[4]More precisely, the number of comparisons.

simple examples that should give you a good intuition for what the asymptotic upper bound really means. Note that for the most part you can determine the asymptotic upper bound of a function by ignoring all but the highest order term in the formula, as well as constant factors.

Here are some examples:

1. If $f(n) = n^2$, then $f(n) = O(n^2)$
2. If $f(n) = 6n^2$, then $f(n) = O(n^2)$
3. If $f(n) = 3n^2 + 2n$, then $f(n) = O(n^2)$
4. If $f(n) = 2n^4 + n$, then $f(n) = O(n^4)$
5. If $f(n) = 2^n$, then $f(n) = O(2^n)$
6. If $f(n) = 2^n + n^5$, then $f(n) = O(2^n)$
7. If $f(n) = 2n^4 + n$, then $f(n) = O(n^{10})$

All of the above, except perhaps the last two examples, can be understood based on the simple explanation we provided earlier about ignoring constants and terms that are not the highest order terms. For example 6, it may not be obvious that n^5 is a lower order term than 2^n, but if you try plotting these two functions on a graph, you will see that 2^n grows much faster than n^5. Example 7 illustrates the fact that the Big-O notation gives an upper bound, not a tight upper bound, so something that is $O(n^4)$ will trivially be $O(n^5)$, $O(n^6)$, $O(2^n)$, etc.

Before we provide the mathematical definition of O, we provide one more motivation for why constants are often ignored. The reason is that we often care more about the rate at which a function grows than the exact amount of time it takes for the algorithm to run on a given input. For example, consider the functions $f_1(n) = 1000n^2$ and $f_2(n) = n^3$, so that $f_1(n) = O(n^2)$ and $f_2(n) = O(n^3)$. The asymptotic running time complexity is much better for f_1, even though for small values of n the function f_1 will actually be more expensive. But we know that for large values of n there will come a point where f_1 performs better than f_2 and we focus on this behavior because we need the better performance exactly when n is large.

Given this understanding of the Big-O notation, the following table summarizes the asymptotic complexity of the two search algorithms and the two sorting algorithms that we covered in this chapter.

Algorithm	Time Complexity
linear search	$O(n)$
binary search	$O(\log_2 n)$
bubblesort	$O(n^2)$
mergesort	$O(n \log_2 n)$

We are finally ready for the formal definition of O, which says that

$$f(n) = O(g(n)) \text{ as } n \to \infty$$

if and only if there exist positive constants c and n_0 such that

$$0 \leq f(n) \leq cg(n) \text{ for all } n \geq n_0 .$$

Example 8.12: Prove that if $f(n) = 3n^2 + 2n$, then $O(f(n)) = n^2$.

Solution: We need to show that for some positive constant c and for some $n \geq n_0$, it is always true that

$$0 \leq 3n^2 + 2n \leq cn^2 .$$

We contend that the values $c = 6$ and $n_0 = 1$ will work. Thus we need to show that

$$0 \leq 3n^2 + 2n \leq 6n^2 .$$

Certainly if we replace $2n$ with $2n^2$ and the new inequality still holds, then the original inequality above holds, since $2n^2 \geq 2n$ when $n > 0$. So, we need to check whether it is always true that

$$0 \leq 3n^2 + 2n^2 \leq 6n^2 .$$

Simplifying, we get:

$$0 \leq 5n^2 \leq 6n^2,$$

which clearly is always true, since $n \geq 1$. □

8.5 Exercises

8.5.1 The *max* function returns the maximum numerical value in a list of integers, such that max({4, 5, 8}) = 8. Describe the algorithm for computing the max function, using a precise English description. Also determine the number of comparisons required by the algorithm in terms of n, the number of items in the list.

8.5.2 The algorithm below is designed to compute the factorial $n!$ of a value n, for any non-negative integer n.

1. $factorial = n$
2. **repeat** as i varies from $n - 1$ down to 1
3. $\quad factorial = factorial \times i$
4. return *factorial*

(a) Given an input value of $n = 4$, show the value of *factorial* after line 3 is executed for each iteration of the **repeat** statement. What is the final value that is returned on line 4 (do not show the value of *factorial* for line 1).

(b) The factorial algorithm is not quite complete and as a result it will not return the correct answer for one non-negative input value. For what *non-negative* input value n will the algorithm not function properly and what value will it return in this case? Finally, how would you modify the algorithm to fix this problem?

8.5.3 Write an algorithm to make a peanut butter and jelly sandwich and then clean up, so that to the extent possible, everything is in the same state as at the start. To help constrain the number of possible solutions and to identify the desired *level* of abstraction, you should use only the functions listed below (but you should use all of them). The final result should be a peanut butter and jelly sandwich on a plate (which in turn is on a table). The relevant objects are: a jar of peanut butter, a jar of jelly, a bag containing a loaf of bread, a knife, a plate, and a table. You should assume that all objects except the table are initially in storage.

$get(x)$:	get object x from storage.
$away(x)$:	put object x back into storage.
$open(x)$:	open x so contents are accessible.
$close(x)$:	close x so contents are not accessible.
$serving(x, y)$:	take one serving from within x and put onto y.
$spread(x)$:	spread what is on the knife onto x.
$top(x, y)$:	put x on top of y.
$clean()$:	clean the knife.

Note: Assume that the *serving* function returns a handle or reference to the serving. Thus $b_1 = serving(bread, table)$ puts a piece of bread onto the table; you can subsequently refer to this piece of bread as b_1. For simplicity, you can ignore the handle for cases where there is no ambiguity.

8.5.4 Use binary search to find the element e in the sorted list given by (a, d, f, g, k, l). For each iteration of the algorithm described on page 260, show the values of *min, max*, and *midpoint* after step 3 of the binary search algorithm completes. Ultimately how many values are compared to "e"? This questions is similar to the question in Example 8.3 and your solution should provide the same information as the solution to that example.

8.5.5 The number of comparisons required by the bubblesort algorithm is $\frac{1}{2}n(n-1)$, so that the algorithm requires $O(n^2)$ comparisons. These values correspond to the algorithm's worst case performance.

(a) For sorting problems, one might expect the best case performance to occur when the input list is already sorted. For our bubblesort algorithm, as described in Section 8.3.2.1, what is the best case performance of the algorithm? Provide your answer in terms of the number of comparisons.

(b) Can you think of a way to improve the best case performance of the algorithm without hurting the worst case performance of the algorithm? In words, explain how.

(c) What is the best case performance, in terms of number of comparisons, with your improvements? How does the number of comparisons grow with n, the size of the list?

(d) Finally, do you think that with your improvements you have produced the optimal best case performance, in terms of number of comparisons and algorithmic complexity? That is, do you think it is impossible to do better? Justify your answer.

8.5.6 The bubblesort algorithm, as described in Section 8.3.2.1, can be implemented in a computer program using the *function* construct. But mathematically speaking, based on what we discussed in Chapter 5, is bubblesort a function? Answer "yes" or "no." Also, which of the properties of functions does it have? State whether it is injective and/or surjective and whether it is invertible. Justify/explain all of your answers.

For simplicity, you may assume that we are sorting integers and the the domain and codomain include all possible lists of integers.

8.5.7 [SN] Table A.1 on page 318 provides the basic information for the people in our sample social network.

(a) If we use the linear search algorithm to search the table for Alyssa, what names (in order) would be checked? How many total name comparisons will be made?

(b) If we use the binary search algorithm to search the table for Alyssa, what values would *min*, *max*, and *midpoint* take after step 3 for each iteration of the algorithm described on page 260?

(c) For the binary search algorithm, what names (in order) would be checked? How many total name comparisons will be made?

(d) (*)In this case which algorithm requires more comparisons? Do these specific results for the search for Alyssa agree with the statement in Section 8.4.3.2 that the binary search algorithm is more efficient than the linear search algorithm? How to you resolve any apparent contradiction?

8.5.8 Similar to the way it was done in Example 8.7, provide a step-by-step description of how mergesort would sort the list of numbers (4 3 1 2) into increasing order. Number each step that involves a *merge* and specify the total number of times a *merge* is done.

8.5.9 Formally prove that if $f(n) = 5n^2 + 3n + 2$ then $f(n)$ is $O(n^2)$

8.5.10 In order to get a good feeling for how different functions grow with increasing n, generate a table that has a column for each of these values of n: (1 10 100 1,000 10,000 100,000). Then, in each row, calculate the value, for each value of n, for each of the following functions: $f_1(n) = n$, $f_2(n) = n \log_2 n$, $f_3(n) = n^2$, and $f_4(n) = 2^n$. There are many algorithms that have time complexities that correspond to each of these four functions.

Chapter 9

Graphs

> At the other end of the spectrum is, for example, graph theory, where the basic object, a graph, can be immediately comprehended.
>
> W. T. Gowers,
> *The Two Cultures of Mathematics*

When you visualize your friends and acquaintances, it is possible you see this in your mind as a collection of people connected to each other by "invisible" links of kinship or friendship. Such pictures seem common and intuitive to many. There is a way to formalize these pictures and to use them to solve problems. These pictures are called *graphs*. By now, you may be familiar with the social network diagram in Figure A.1 of the Appendix, which has formed the basis of many examples and exercises in prior chapters. This diagram of a social network is a graph in the sense we are introducing here.

Let's consider some of the useful things that representing a problem as a graph will allow us to do. Look closely at the portion of the graph in Figure A.1 that contains the social connections of Ellen, Alyssa, Lauren, and Grace. These people are each represented as "places" or *vertices* in the graph and are reflected as a circle each containing their respective names as shown in Figure A.1. Each vertex has some lines connecting it to other vertices, representing the connection of this person to other people in the social network. These lines are called *edges*. The pattern of connections between the vertices allows us to see immediately that in this example, three of the four (i.e., Ellen, Alyssa and Lauren) all know each other, a fact that a savvy social networking site might be able to use. Secondly, we can see that Alyssa and Ellen both know Grace—but Lauren doesn't. Given that her friends Allysa and Ellen both like Grace, perhaps Lauren would like her too, another fact that a savvy social network site could use!

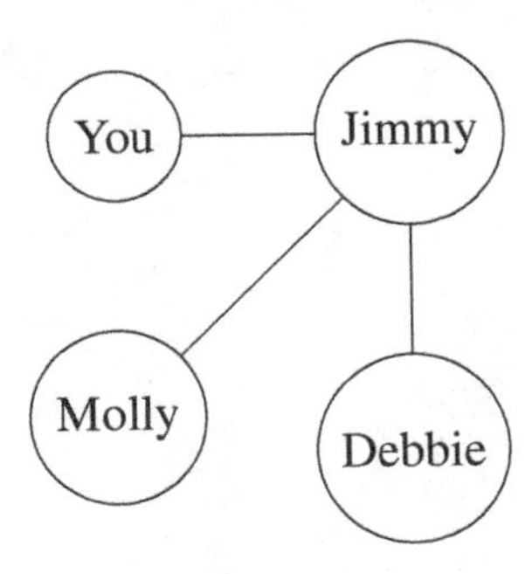

Figure 9.1: Graph representing Jimmy's siblings

Another useful problem we can address with a graph is working out who is related to whom on a social networking site. Let's say you declare one of your social network friends, Jimmy, to be your brother. We can visualize this as a graph with two vertices: one for you and one for Jimmy. The vertices are connected by an edge, linking them together. Jimmy might list all three of his siblings, one of whom is you, and the other two are his sisters Molly and Debbie. Visualize this as two more vertices on the graph, connected by edges to Jimmy (see Figure 9.1). But this graph of sibling connections makes it very easy to determine (by following the lines) that Molly and Debbie are your sisters as well, something you have not shared with your social networking site. This discovery is something that you're quite happy about, however, when Molly's birthday rolls around and you are prepared, thanks to an advance hint from your social networking site!

This novel way of solving problems was invented by a Swiss mathematician, *Leonhard Euler* (pronounced "Oiler"). Euler was considering the layout of the seven bridges in the city Königsberg (then in Prussia, but now called Kaliningrad and in Russia). The city was unusual in that it was built on an island in the River Pregel, as shown in Figure 9.2. The problem Euler was considering was to determine if it was possible for one to walk over all seven bridges exactly once and end up back where one started.

Euler had the idea of modifying the picture in Figure 9.2 so that it just described the connections between the four land masses involved (which we can think of as A, B, C and D in Figure 9.2). His modified picture is reflected in Figure 9.3.

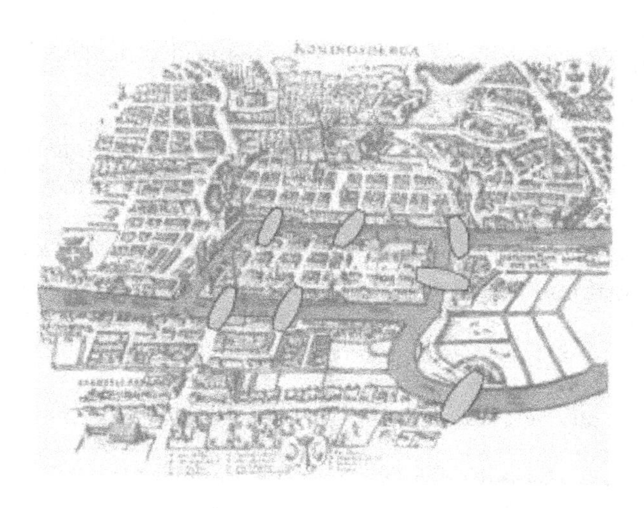

Figure 9.2: The bridges of Königsberg

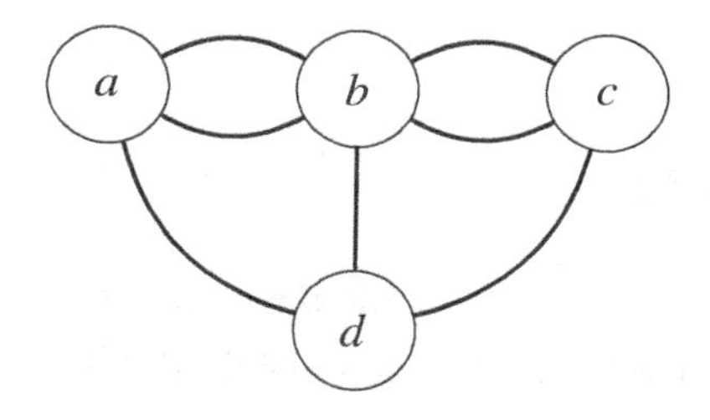

Figure 9.3: Graphical representation of the bridges of Königsberg

From this visual representation of the problem, Euler was able to show that it is impossible to traverse all the bridges exactly once and end up where one started. We will see how he did this later in, in Section 9.2.2.

It turns out that there are many problems that can be better visualized in this way using graphs, for example:

- People in a social networking website: each Facebook or MySpace profile can be viewed as a vertex in a graph; the links to friends and others are the edges in the graph.

- Cities in a country: the vertices are the cities and the edges are the intercity highways that link the cities.

- Jobs in a "to-do" list: the vertices are the jobs and the edges are the dependencies between the jobs (e.g., read the assigned homework chapter *before* doing the assigned homework questions.)

- Electrical connections: The vertices are the electrical appliances and the edges are the copper wires between appliances.

- The Internet graph: The vertices are devices on the Internet, and the edges are simply the direct connections between devices.

Once a problem has been expressed as a graph, then there are a number of useful questions that graph theory can help answer:

- Is it possible to visit every vertex and end up back where one started?

- Is there any vertex that cannot be reached from other places?

- What is the shortest distance between any two vertices?

- How can all the vertices be connected together using the fewest number of edges?

9.1 Graph Notation

In this section, we will introduce some basic notation that will allow us to talk about graphs and solving problems using graphs.

9.1.1 Vertices and Edges

A graph is a set V of *vertices* (or *nodes*) that are joined by a set E of *edges*. Each vertex is given a label or name, so the set V is written as a set of vertex labels. The edges, that is, the "lines" joining the vertices, can also be labeled or named, so the set E is written as a set of edge labels.

Example 9.1: Write down the set of edges and the set of vertices in the graph G_1 shown in Figure 9.4.

Solution: In the graph G_1 in Figure 9.4, the set V of vertices is $V = \{a, b, c\}$ and the set E of edges is $E = \{e_1, e_2, e_3\}$. □

To show that a set of vertices and a set of edges is always needed to completely specify a graph, we will write the graph G as $G = (V, E)$. Sometimes it is convenient to avoid labeling the edges. In that case, the edges are specified by saying which vertices are connected by an edge. For example edge e_1 in the graph in Figure 9.4 joins vertices a and c, so we can just as easily write the set $\{a, c\}$ to indicate the edge between a and c.

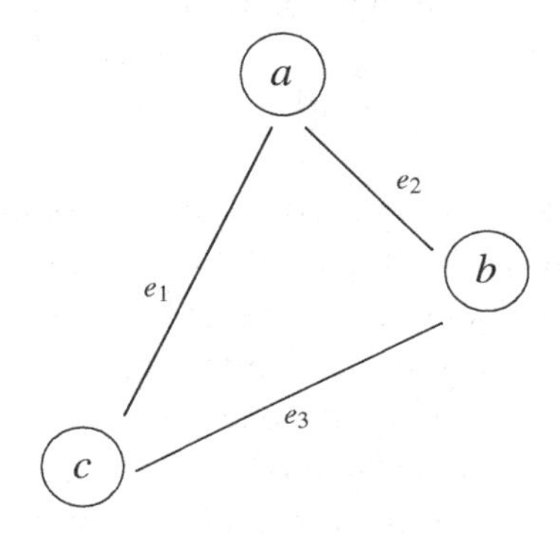

Figure 9.4: Graph for Example 9.1

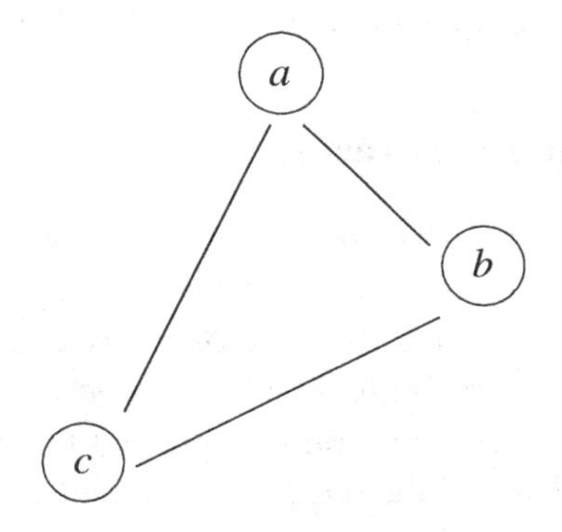

Figure 9.5: Graph for Example 9.2

Example 9.2: Write down the set of edges and the set of vertices in G_2 shown in Figure 9.5.

Solution: In the graph G_2 in Figure 9.5, the set of vertices is $V = \{a, b, c\}$, but the set of (unlabeled) edges is now $E = \{\{a, b\}, \{b, c\}, \{a, c\}\}$. □

Remark. Representing the edges E as a set of (unordered) pairs of vertices leads to one problem: you can have at most one edge between any pair of vertices! To see this, look once again at the Königsberg graph in Figure 9.3. Since there are multiple edges from a to b and from c to d, we would write the edge set as

$$\{\{a, b\}, \{a, b\}, \{b, c\}, \{b, c\}, \{a, d\}, \{c, d\}\}.$$

But this is an improper representation of the set

$$\{\{a, b\}, \{b, c\}, \{a, d\}, \{c, d\}\},$$

and so it does *not* represent the edge set of the Königsberg graph.

The problem here is that Figure 9.3 is not a graph, strictly speaking, since some pairs of vertices are connected by more than one edge. Figure 9.3 is actually a *multigraph*, which is simply a graph allowing pairs of nodes to be connected by more than one edge.

It is possible to extend the "pair of vertices" edge representation found in graphs, so that this representation applies to multigraphs as well. However, we're going to omit the tedious details.

All the results of this chapter apply to both graphs and multigraphs. For the sake of conciseness, we will be using the term "graph" to include both graphs and multigraphs in the material that follows. □

9.1.2 Directed and Undirected Graphs

A graph is called a *directed graph* (or *digraph*, for short) if the edge from vertex v_1 to vertex v_2 can only be traveled from v_1 to v_2 and *not* from v_2 to v_1. A directed edge represents a one-way constraint. For example, if some of the Königsberg bridges only allowed one-way traffic, then we would have to use a directed graph to represent the problem. If the graph represented a to-do list and the edges represented what jobs must be done before other jobs, then a directed graph would also have to be used. In a social network, when a person issues a *friend* request, this is a one way request and also will need to be represented using a directed graph. (Consider the one-way arrows shown in Figure A.1 of the Appendix.)

The graphs that we have looked at in previous examples have been examples of *undirected graphs*.

We run into a problem if we try to represent the edges in a directed graph using the method that we have used for undirected graphs. For example, the directed edge from vertex v_1 to vertex v_2 cannot be written as $\{v_1, v_2\}$, since the set $\{v_1, v_2\}$ is the same as the set $\{v_2, v_1\}$, but the directed edge from v_1 to v_2 is not the same as the edge from v_2 to v_1. The solution to this problem is to represent the directed edge as an ordered pair (v_1, v_2), since (v_1, v_2) is *not* the same as (v_2, v_1).

Example 9.3: List the set of edges and the set of vertices in the graph G_3 shown in Figure 9.6.

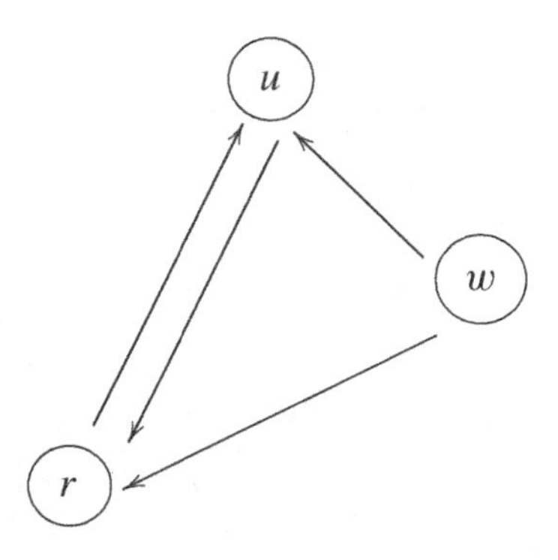

Figure 9.6: Directed graph for Example 9.3

Solution: The graph $G_3 = (V, E)$ for the directed graph shown in Figure 9.6 has the vertex set

$$V = \{u, r, w\}$$

and the edge set

$$E = \{(r, u), (u, r), (w, r), (w, u)\}.$$ □

9.1.3 Complete Graphs

A graph is said to be *complete* if there is an edge connecting every pair of vertices. For instance, an undirected graph having two vertices is complete if there is an edge between the vertices, e.g., we have $G = (V, E)$ with $V = \{a, b\}$ and $E = \{\{a, b\}\}$. With three nodes, there would have to be three edges (such as the graph in Figure 9.5). How many edges should there be for a complete undirected graph with four vertices? What about five?

A directed graph with two vertices v_1 and v_2 must have two edges (v_1, v_2) and (v_2, v_1) to be complete. How many edges should there be for a complete directed graph with three vertices? What about four and five vertices?

Example 9.4: Answer the following questions about the graph shown in Figure 9.7:

(a) How many vertices are there and how many edges are there?

(b) Is this a directed or an undirected graph?

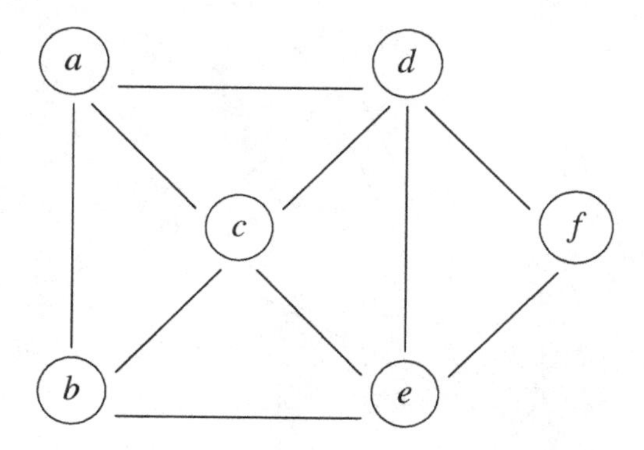

Figure 9.7: Graph for Example 9.4

(c) Is this a complete graph?

(d) What is in the set V of vertices and the set E of edges?

Solution:

(a) There are six vertices and there are ten edges.

(b) It is an undirected graph. There are no arrows on the edges.

(c) It is not a complete graph; vertex f is not directly connected to vertex c for example.

(d) We have

$$V = \{a, b, c, d, e, f\},$$

and

$$E = \{\{a, b\}, \{a, c\}, \{a, d\}, \{d, c\}, \{d, e\}, \{d, f\}\{c, b\}, \{c, e\}, \{f, e\}, \{b, e\}\}$$

□

Example 9.5: [SN] A *subgraph* of a graph G is a graph that just consists of only some of the vertices and edges from G. We will expand upon this in a later section. Consider the subgraph of Figure A.1 that contains only the vertices Niko, Sam, Frank and Luka. This subgraph contains only those edges from the graph that are between these vertices. Answer the following questions about this subgraph:

(a) How many vertices are there?

(b) How many edges are there?

(c) Is this a directed or an undirected graph?

(d) Is this a complete graph?

(e) What is in the set V of vertices and the set E of edges.

Solution:

(a) There are four vertices.

(b) There are six edges.

(c) This is an undirected graph (no arrows).

(d) This is a complete graph.

(e) We have

$$V = \{\text{Niko, Sam, Frank, Luka}\}$$

and

$$E = \{\{\text{Niko, Sam}\}, \{\text{Sam, Luka}\}, \{\text{Luka, Frank}\}, \{\text{Frank, Niko}\}, \{\text{Niko, Luka}\}, \{\text{Sam, Frank}\}\}$$ □

9.2 Euler Trails and Circuits

Recall that the problem Euler was addressing at Königsberg was to see if each of the seven bridges could be traversed just once and end up back at the starting point. In the next subsections we will introduce some additional terminology that allows us to address this problem.

9.2.1 Walks, Trails, Circuits and Cycles

A *walk* is any sequence of vertices and edges starting at some vertex v_0 and ending at a vertex v_n. A *trail* is a walk in which no edge is traversed more than once. In contrast, a *path* is a walk in which no vertex is traversed more than once. A *circuit* is a trail that begins and ends at the same vertex. Finally, a *cycle* is any circuit in which the start vertex is the *only* vertex that appears more than once.

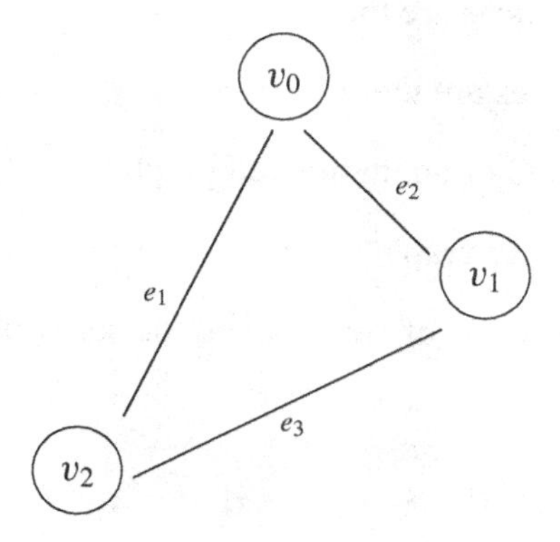

Figure 9.8: Graph for Examples 9.6 and 9.7

Example 9.6: How many trails are there from v_0 to v_2 in the graph shown in Figure 9.8?

Solution: There are two trails from v_0 to v_2 in the graph shown in Figure 9.8:

$$w_1 = (v_0, e_2, v_1, e_3, v_2)$$
$$w_2 = (v_0, e_1, v_2).$$

□

Example 9.7: Consider the following four sequences of vertices and edges of the graph in Figure 9.8. Indicate for each sequence whether it is a walk, a trail, a path, a circuit, a cycle, or none of these. Explain your answer in each case.

(a) $(v_0, e_3, v_1, e_1, v_2)$.

(b) (v_0, e_1, v_2, e_3).

(c) $(v_1, e_3, v_2, e_3, v_1)$.

(d) $(v_2, e_3, v_1, e_2, v_0, e_1, v_2)$.

Solution:

(a) A walk is any sequence of vertices and edges that starts at one vertex and ends at another, so the edge following a vertex must be an edge connected to that vertex. Here we see that e_3 follows v_0 but e_3 is not connected to v_0. Hence this is not a walk.

(b) This sequence does not end with a vertex and hence can't be a walk.

(c) This sequence is a walk, but is not a trail since the edge e_3 is repeated, and it is not a path since v_1 is repeated. Even though it it starts and ends with the same vertex, it is not a circuit since it is not a trail.

(d) This is a walk, a trail (no repeated edges), and since it starts and ends with v_2 it is also a circuit. Since v_2 is the only repeated vertex, this is also a cycle. □

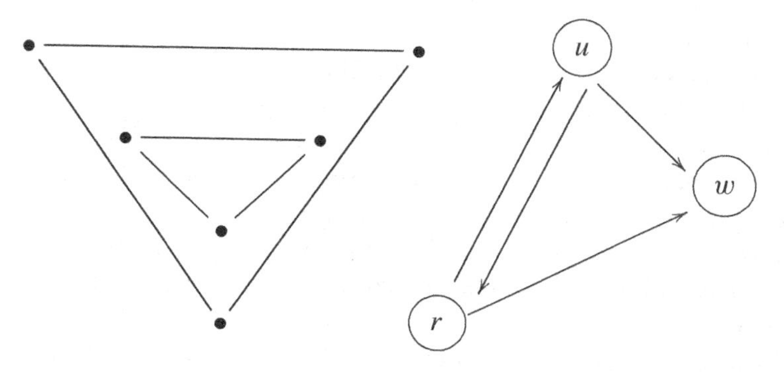

Figure 9.9: Are these graphs connected?

A graph is *connected* if there is a walk from any vertex v_1 to any other vertex v_2. It is easy to see if an undirected graph is not connected, since one or more of the vertices will be completely separate from the others. It's harder to tell if a directed graph is not connected. For example, look at the two graphs in Figure 9.9. Neither of these graphs is connected. However, it's much easier to check this fact for the undirected graph on the left than it is for the directed graph on the right.

Finally, an *Euler trail* is a trail that includes every edge in the graph. An *Euler circuit*, as you might guess, is an Euler trail that begins and ends at the same vertex.

9.2.2 When Can We Find Euler Trails and Circuits?

When Euler was looking at the Königsberg bridge problem, what he was actually doing, we would say today, was checking to see if an Euler circuit existed for the graph in Figure 9.3. One way to address this is to try to find every possible circuit, one by one, and check if it is an Euler circuit. This can take a long time of course, and if you do not find an Euler circuit it doesn't prove one doesn't exist: it maybe that one exists and it is one that you haven't tried yet!

As it happens, there is an easy way to check if an Euler circuit exists for a connected graph. Each edge in a circuit can only be traversed once, and so every vertex in an Euler circuit has to have an exit edge (a "way out") for *every* entry edge (a "way in"). If this were not the case, then you could enter a vertex during your walk and be stuck there! So every vertex has to have two edges (one in and one out) or four edges (two in and two out) or six edges (three in and three out), and so forth. In other words, no vertex can have an odd number of edges, since we would wind up getting stuck in that vertex.

The number of edges that a vertex has is referred to as the *degree* of the vertex. So, by our reasoning, if any vertex is of odd degree, then we can be sure that the graph does not have an Euler circuit. It is possible (but somewhat harder) to show that the converse is also true: if all the vertices have even degree, than an Euler circuit does exist.

Now we can duplicate Euler's results for the Königsberg problem: all the vertices in Figure 9.3 have odd degree, therefore an Euler circuit is not possible!

What can we say about Euler trails? We only need to talk about Euler trails that are *not* Euler circuits, since we've just finished talking about Euler circuits. So suppose that a graph $G = (V, E)$ has an Euler trail that's not an Euler circuit. Such an Euler trail has a starting vertex (call it v) and an ending vertex (call it w). We artificially add a new edge, connecting w to v, giving us a new graph $G' = (V, E')$. Consider a trail consisting of all the edges in the original Euler circuit from v to w, followed by the newly-added edge from w to v. This is an Euler circuit in the graph G'. Since G' has an Euler circuit, the G'-degrees of all the vertices in V are even. What does this say about the G-degree?

- For any vertex other than v or w, the G-degree is the same as the G'-degree, which is even.

- For v and w, the G-degree is one less than the G'-degree, which means that v and w have odd degree in the original graph.

In short, we have shown that if the graph has an Euler trail that is not an Euler circuit, then exactly two vertices have odd degree, with the rest having even degree. Moreover, a slight variation of the "add an extra edge" trick shows that the converse is true, i.e., if exactly two vertices have odd degree, then there is an Euler trail for which these two vertices are the initial and terminal vertices.

In summary, we have the following

Fact. *Let $G = (V, E)$ be a graph.*

- *If every vertex in V has even degree, then G has an Euler circuit.*
- *If exactly two vertices $v, w \in V$ have odd degree, then G has an Euler trail, starting at v and ending at w. However, G does not have an Euler circuit.*
- *In all other cases, G has neither an Euler trail nor an Euler circuit.*

□

Example 9.8: Does the graph shown in Figure 9.7 have an Euler trail?

Solution: Examining each of the vertices in the graph shown in Figure 9.7 shows that one vertex has degree 2 (that is, f), two vertices have degree 3 (that is, a and b) and three vertices have degree 4 (that is, c, d and e). There are two vertices of odd degree, so there is an Euler trail. A little experimentation will yield the trail $(a, b, c, d, e, f, d, a, e, b)$. The trail begins and ends with a vertex of odd degree as must be the case if there are two vertices of odd degree. □

9.3 Weighted Graphs

It is sometimes natural to associate a number with each edge on a graph. For example, a graph that shows the cities in the northeast of the United States as vertices and the interstate highways between them as edges might also include the distance between cities on each edge. This number associated with an edge is called the edge *weight* and a graph that has weights on every edge is called a *weighted graph* (see Figure 9.10).

Additional examples of weighted graphs include the following:

- A graph that shows the electrical connections between appliances in a building and that includes as weights the length of copper wire used on each edge.
- A graph that shows the cell phone communication towers in a locality and that has weights that capture the cost of sending a message from one tower to another.

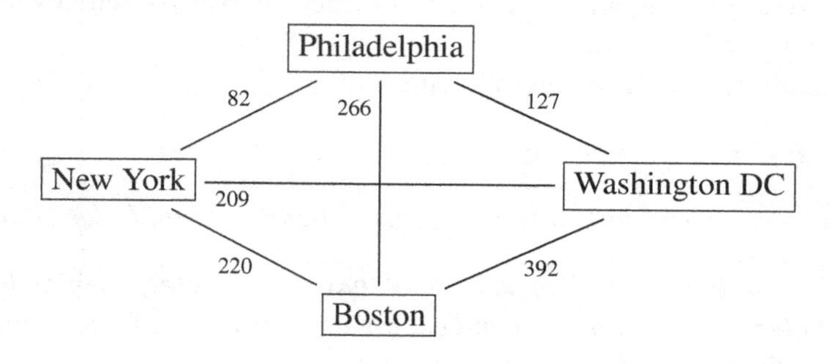

Figure 9.10: Weighted graph of U.S. cities

A *weighted* graph G is given by $G = (V, E, W)$, where V is a set of vertices, E is a set of edges (as before) and W is a set of weights. This set W of weights is a set of ordered pairs (e, w). The first component of each pair is the edge e and the second is the weight w. Of course, each edge e between vertices v_1 and v_2 is also either a set (undirected graph) or an ordered pair (directed graph). So the ordered pairs in W look like $(\{v_1, v_2\}, w)$ for an undirected graph and $((v_1, v_2), w)$ for a directed graph.

Example 9.9: Write down the description of the weighted graph in Figure 9.10 in terms of sets of vertices V, edges E, and weights W associated with each edge.

Solution: To save a little time and space, let us write abbreviate "Boston", "New York", "Philadelphia", and "Washington DC" as b, n, p, and w. We find that the weighted graph $G = (V, E, W)$ in Figure 9.10 can be written as follows:

$$\begin{aligned} V &= \{b, n, p, w\}, \\ E &= \{\{n, p\}, \{p, w\}, \{n, w\}, \{n, b\}, \{b, p\}, \{b, w\}\} \\ W &= \{(\{n, p\}, 82), (\{p, w\}, 127), (\{n, w\}, 209), \\ &\quad (\{n, b\}, 220), (\{b, p\}, 266), (\{b, w\}, 392)\}. \end{aligned}$$

□

The *weight* w_G of a graph G is just the sum of all the weights on all the edges in the graph G. If the weight on any edge e is written as w_e then the weight of the graph can be written using summation notation as

$$w_G = \sum_{e \in E} w_e.$$

For example, the weight of the graph in Example 9.9 is calculated as

$$w_G = 82 + 127 + 209 + 220 + 266 + 392 = 1,296.$$

The weight of a walk in a graph G is just the sum of the weights along the edges of the walk. For the graph in Example 9.9 the weight of the walk $n \to p \to w$ from New York through Philadelphia to Washington DC is

$$w_{n \to p \to w} = 82 + 127 = 209.$$

Example 9.10: Calculate the weights of the following walks using the graph in Figure 9.10:

(a) New York to Washington DC via Boston.

(b) Boston to Washington DC via Philadelphia.

(c) The return trip from Washington DC to Boston via Philadelphia and New York.

Solution: The weights are as follows:

(a) $w_{n \to b \to w} = 220 + 392 = 612.$

(b) $w_{b \to p \to w} = 226 + 127 = 353.$

(c) $w_{w \to p \to n \to b} = 127 + 82 + 220 = 429.$ □

9.4 Minimum Spanning Tree

Another useful question that graph theory can answer is how to select the minimum amount of some resource connecting a set of objects. For example, if a building is being equipped with fiber-optic cable, then to minimize the cost of the job, the installer would like to use the smallest amount of cable that connects all the locations together.

Consider the graph in Figure 9.11. This graph shows locations in a building (the vertices) to which fiber-optic communications needs to be provided (e.g., for video, phone, or network connections). The edges are the fiber-optic connections and the weights are the amount of fiber for each connection. Note that there are several ways for a message to travel along the fiber from a to d: directly from a to d (with weight 16), or another choice is from a through c to d (with total weight $21 + 11 = 32$), and so forth. Any installer who wanted to make a decent profit on the installation

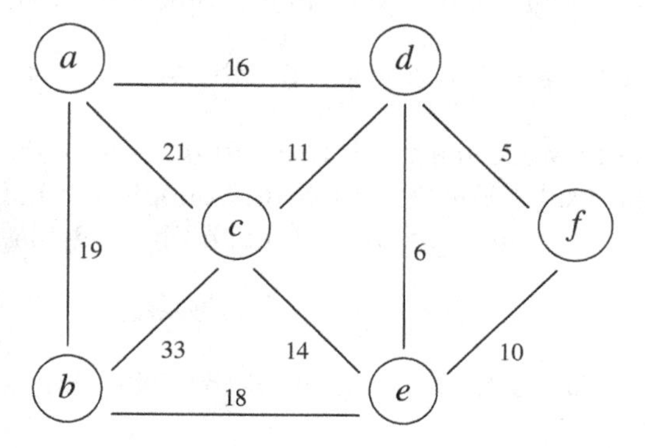

Figure 9.11: Fiber-optic connections in building

would like to connect all the locations (vertices) in such a way that no extra routes exist between vertices (so that no extra fiber is used) and that for each pair of vertices only the smallest weight edge is left (that is, the one that used up the smallest amount of fiber). To solve this problem, we need to introduce the idea of a minimum spanning tree.

9.4.1 Spanning Trees

As we saw in Example 9.5, a *subgraph* of a graph G is a graph that just consists of some of the vertices and edges from G. A *spanning tree* of a connected graph G is a special kind of subgraph of G that includes all the vertices that are in G (but not necessarily all the edges) and in which there are no cycles possible. Note that there are many cycles in the graph in Figure 9.11. For example, one cycle connects the vertices a, b and c, while another connects the vertices d, c, and e; you should be able to find others.

Example 9.11: Figure 9.12 is a completely connected undirected graph with three nodes. How many spanning trees can we make from this graph?

Solution: We can make spanning trees by deleting any edges that cause cycles, but that leave a connected graph. There are three edges in the complete graph G in Figure 9.12 and they all are part of the one cycle (or loop) in that graph. Thus we can get three spanning trees for this graph, each made by deleting one of the edges in G, as is shown in Figure 9.13. □

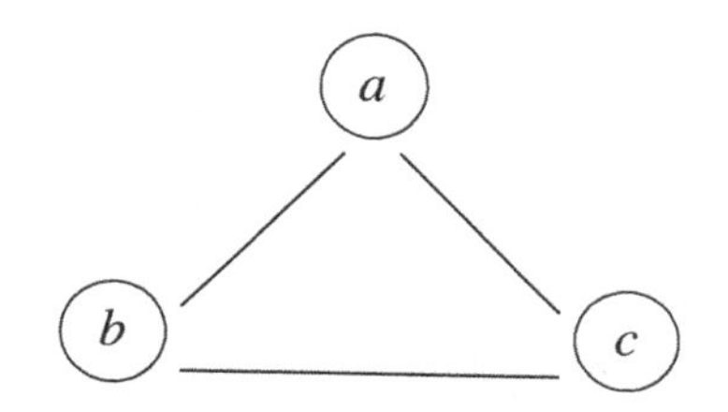

Figure 9.12: Graph for Example 9.11

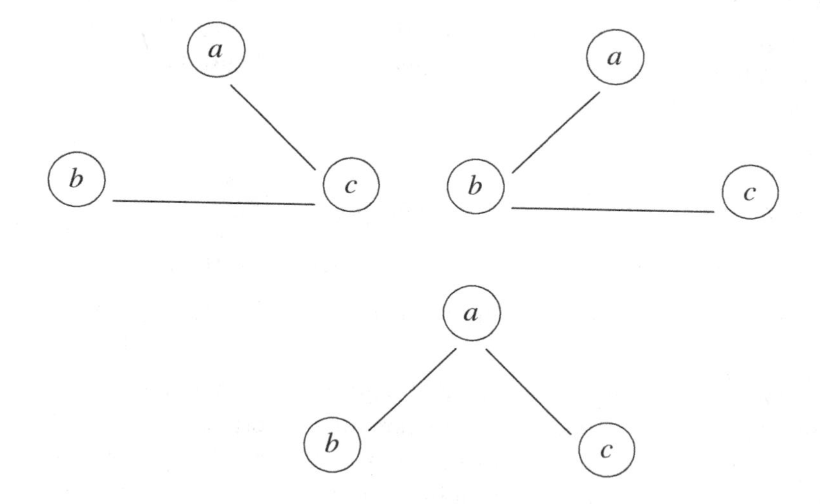

Figure 9.13: Spanning trees for Figure 9.12

9.4.2 Prim's Algorithm for the Minimum Spanning Tree

A *minimum spanning tree* (or MST) of a graph G is a spanning tree of G that has the least weight of all the spanning trees of G. The MST for the graph in Figure 9.11 is what the prudent installer would use to minimize costs. One approach for computing the MST is to write out, one by one, all of the spanning trees of Figure 9.13, and then calculate the weight of each one. That could take a while for large graphs! It turns out there is an easier solution to the problem. This is called *Prim's* MST *algorithm* and was developed by R. C. Prim in 1957. This algorithm[1] consists of the

[1]Recall that an algorithm is a sequence of concrete instructions to solve a problem (see Chapter 8).

following steps:

1. Write down all the edges in the graph in order, from smallest to biggest weight.
2. Working from the smallest to the biggest, add each edge into MST only if it does not make a cycle.

We *stop* when all the vertices are in the graph *and* the graph is connected (i.e., no isolated vertices).

Example 9.12: Use Prim's algorithm to calculate the MST for the graph shown in Figure 9.11 of fiber-optic connections in a building. The weight of an edge in this graph is the amount of fiber-optic cable used in the connection represented by the edge.

Solution:

(a) We first list the edges in order of ascending weight:

$$(\{d, f\}, 5), (\{d, e\}, 6), (\{f, e\}, 10), (\{d, c\}, 11), (\{c, e\}, 14)$$
$$(\{a, d\}, 16), (\{b, e\}, 18), (\{a, b\}, 19), (\{a, c\}, 21), (\{b, c\}, 33)$$

(b) We then add the edges into the MST starting at the smallest, unless the edge makes a loop. We stop when all the vertices are present. Figure 9.14 shows the first three edges considered for the MST in order, from left to right. The third edge is rejected from the MST because it would make a loop! We ignore this edge and move to the fourth edge in order $\{d, c\}$ and add that in. We next reject $\{c, e\}$ however because it also would make a loop. We add $\{a, d\}$ and $\{b, e\}$. Since that includes every vertex, we're done! The complete MST is shown in Figure 9.15. As you can easily see it includes all the vertices and uses far less fiber-optic cable than the original. The weight of the MST is $5 + 6 + 11 + 16 + 18 = 56$. □

9.5 Matrix Notation For Graphs

Let's reflect for a moment on how we've been describing graphs so far:

- by a picture, and
- by explicitly listing vertices and edges.

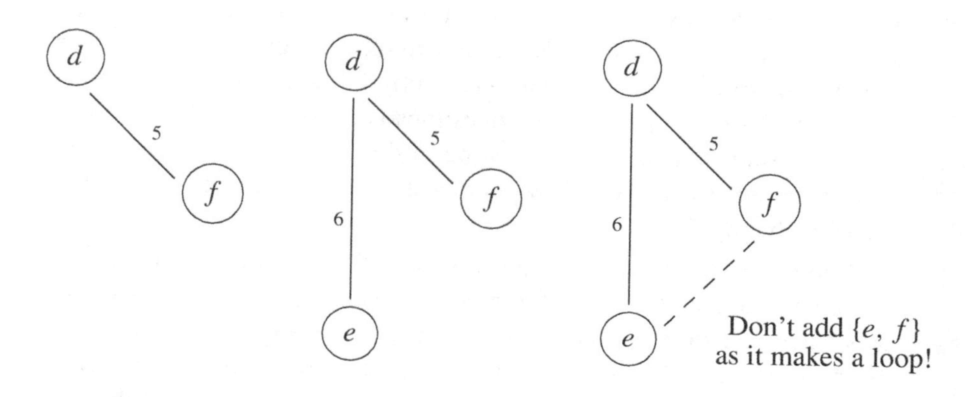

Figure 9.14: First steps in construction of MST for Figure 9.11

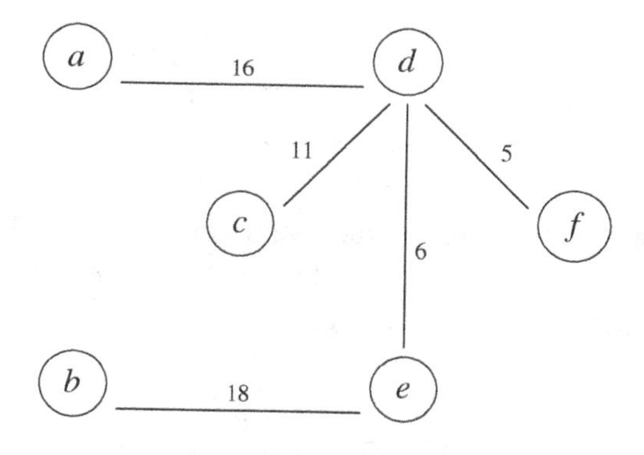

Figure 9.15: MST for Figure 9.11

If the graph is small, this is perfectly fine. But these techniques don't work very well for large graphs. What do we mean by "large" here? Let's go back to the Facebook example mentioned at the beginning of the chapter. As of the writing of this book, Facebook had over 400 million active users, with the average user having 130 friends.[2] Suppose that we're less ambitious, and want to deal with graphs of only moderate size (say, a few dozen nodes); we would probably want to use a computer to answer any questions (such as finding shortest distances or minimal spanning trees) we might have. But how would we input a graph into the computer? We could

[2]See `http://www.facebook.com/press/info.php?statistics` for the most up-to-date statistics.

certainly scan a drawing of the graph, storing it in some graphical format, but computers are not yet as good at image recognition as human beings; having the computer translate that JPG or PDF file into a form that it could use for answering graph-theoretic questions is no simple task. Moreover, the sheer tediousness of typing a listing of dozens of vertices and (potentially) hundreds of edges almost guarantees that we'll make a mistake in preparing the input.

It turns out that matrix notation fits the bill here. A *matrix* is nothing more than a rectangular array of numbers; you may have learned about matrices when you studied high school math. But there will be an important difference between the matrices you may have seen before and the matrices that we'll be discussing. Your earlier exposure to matrices would have been to matrices whose entries could be any numbers whatsoever, such as

$$\begin{bmatrix} 2 & 3 & -1 & 5 \\ 7.01 & -4.2 & 13.77 & 14.92 \\ -6 & 42 & 1.7 & 0.5 \\ 0 & 19 & -3.14 & 2 \end{bmatrix}.$$

You learned various operations on such matrices, such as addition and multiplication. In this chapter, we will be looking at *Boolean matrices*, whose entries are only zeros and ones, such as

$$\begin{bmatrix} 0 & 1 & 1 & 1 \\ 1 & 0 & 1 & 0 \\ 0 & 0 & 1 & 1 \\ 1 & 1 & 0 & 0 \end{bmatrix}.$$

We will look at operations such as the Boolean sum and Boolean product; by the end of this section, we hope to have convinced you of their usefulness.

Let's start out with the graph given by Figure 9.16, which is similar to Figure 9.7. However, in this graph we are labeling the vertices with numbers rather than letters; moreover, this graph is directed, whereas the graph represented in Figure 9.7 is undirected.

We can represent this graph by a table, in which the rows and columns are headed by vertex numbers. We then put a T into a particular table entry if there is an edge going from the node corresponding to the row number to the node corresponding to the column number. Thus we get:

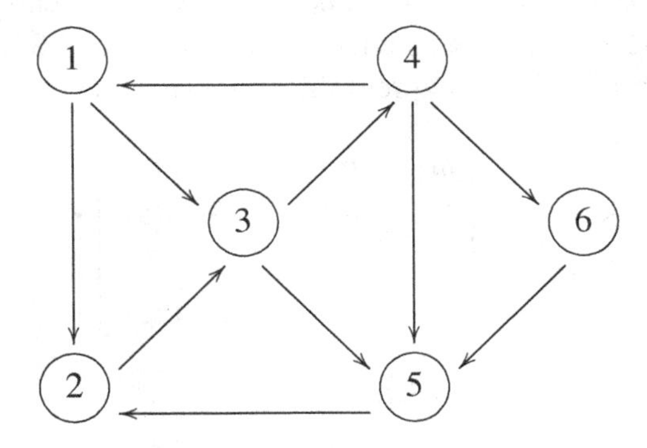

Figure 9.16: A directed graph

	1	2	3	4	5	6
1	F	T	T	F	F	F
2	F	F	T	F	F	F
3	F	F	F	T	T	F
4	T	F	F	F	T	T
5	F	T	F	F	F	F
6	F	F	F	F	T	F

Look at row 1 of this table. There is a T in the second and third positions of this row, because there are edges $(1, 2)$ and $(1, 3)$ in the graph. Each other T that appears can be explained analogously. Similarly, there is an F in the first, fourth, fifth, and sixth positions of the first row because there is *not* an edge going from node 1 to nodes 1, 4, 5, or 6.

Next we throw away the row and column headers; we also replace each F by a 0 and each T by a 1. This gives us the matrix

$$\begin{bmatrix} 0 & 1 & 1 & 0 & 0 & 0 \\ 0 & 0 & 1 & 0 & 0 & 0 \\ 0 & 0 & 0 & 1 & 1 & 0 \\ 1 & 0 & 0 & 0 & 1 & 1 \\ 0 & 1 & 0 & 0 & 0 & 0 \\ 0 & 0 & 0 & 0 & 1 & 0 \end{bmatrix}$$

of the graph in Figure 9.16.

Here's how this process works in general. Let $G = (V, E)$ be a graph (either directed or undirected). Suppose that $V = \{v_1, v_2, \ldots, v_n\}$, so that we have chosen a particular ordering of the vertices. The *adjacency matrix* of G is an $n \times n$ array of zeros and ones, with the entry in row i and column j being 1 if there is an edge from v_i to v_j, and 0 if there is not an edge from v_i to v_j. We can express this more succinctly by writing

$$M = \begin{bmatrix} m_{1,1} & m_{1,2} & \dots & m_{1,n} \\ m_{2,1} & m_{2,2} & \dots & m_{2,n} \\ \vdots & \vdots & \ddots & \vdots \\ m_{n,1} & m_{n,2} & \dots & m_{n,n} \end{bmatrix},$$

where for each $i, j \in \{1, \dots, m\}$, we have

$$m_{i,j} = \begin{cases} 1 & \text{if there is an edge from } v_i \text{ to } v_j, \\ 0 & \text{otherwise.} \end{cases}$$

Note that the entries of an adjacency matrix are zeros and ones, i.e., an adjacency matrix is Boolean. Now a 0 in a particular position in an adjacency matrix means that there's no edge between the two vertices indicated by that position, whereas a 1 means that there *is* a direct connection. So we can think of 0 and 1 as representing the logical values F and T, answering the question "is there an edge between two vertices?" Going back to Chapter 3, perhaps you can now appreciate why we call these matrices "Boolean."

Now if we only wanted to represent a graph, a table would be fine. However, using a matrix to represent a graph allows us to perform various operations on the matrix, which can be used to answer questions about the graph. For example, suppose we know that two people are on Facebook. They might be friends, or friends of friends, or friends of friends of friends (and so forth). Let's say that two Facebook members are *connected* if there is such a "chain of friends" connecting them. For example, suppose that the Facebook graph looked like Figure 9.16.

- The graph gives us direct connections from 1 to 2 and from 1 to 3.
- 1 is connected to 4, since there is a friendship chain $1 \to 3 \to 4$.
- 1 is connected to 5, since there is a friendship chain $1 \to 3 \to 5$.
- 1 is connected to 6, since there is a friendship chain $1 \to 3 \to 4 \to 6$.

Of course, this is a fairly ad hoc way to figure out connectedness. Is there a more systematic way to find out whether two Facebook members are connected? To answer this question, we need to discuss two matrix operations, namely Boolean sum and Boolean product. The first is fairly straightforward; the second is less so.

Let $A = [a_{i,j}]_{1 \le i,j \le n}$ and $B = [b_{i,j}]_{1 \le i,j \le n}$ be $n \times n$ Boolean matrices.

- The *Boolean sum* $A \vee B$ of A and B is the $n \times n$ Boolean matrix $C = [c_{i,j}]_{1 \le i,j \le n}$, with

$$c_{i,j} = a_{i,j} \vee b_{i,j} \qquad (1 \le i, j \le n)$$

  Note that we use the logical disjunction ("or") symbol $\vee$ to denote the Boolean sum matrix operation. This should remind you that the Boolean sum of two matrices is given by a component-wise logical disjunction.

- The *Boolean product* $A * B$ of A and B is the $n \times n$ Boolean matrix $C = [c_{i,j}]_{1 \le i,j \le n}$, with

$$c_{i,j} = (a_{i,1} \wedge b_{1,j}) \vee (a_{i,2} \wedge b_{2,j}) \vee \cdots \vee (a_{i,n} \wedge b_{n,j}) \qquad (1 \le i, j \le n)$$

This formula may be intimidating but an example will show that computing the Boolean product is quite simple.

Example 9.13: Let

$$A = \begin{bmatrix} 0 & 1 & 0 \\ 1 & 1 & 0 \\ 0 & 1 & 1 \end{bmatrix} \qquad \text{and} \qquad B = \begin{bmatrix} 1 & 0 & 0 \\ 0 & 1 & 1 \\ 1 & 0 & 0 \end{bmatrix}$$

Calculate the Boolean sum and Boolean product of A and B.

Solution: We have

$$A \vee B = \begin{bmatrix} 1 & 1 & 0 \\ 1 & 1 & 1 \\ 1 & 1 & 1 \end{bmatrix} \qquad \text{and} \qquad A * B = \begin{bmatrix} 0 & 1 & 1 \\ 1 & 1 & 1 \\ 1 & 1 & 1 \end{bmatrix}$$

Since the Boolean sum is simpler than the Boolean product, let's just look at the latter. Let $C = A * B$. Then

$$\begin{aligned} c_{1,1} &= (a_{1,1} \wedge b_{1,1}) \vee (a_{1,2} \wedge b_{2,1}) \vee (a_{1,3} \wedge b_{3,1}) \\ &= (0 \wedge 1) \vee (1 \wedge 0) \vee (0 \wedge 1) \\ &= 0 \vee 0 \vee 0 = 0, \\ c_{1,2} &= (a_{1,1} \wedge b_{1,2}) \vee (a_{1,2} \wedge b_{2,2}) \vee (a_{1,3} \wedge b_{3,2}) \\ &= (0 \wedge 0) \vee (1 \wedge 1) \vee (0 \wedge 0) \\ &= 0 \vee 1 \vee 0 = 1, \end{aligned}$$

and so forth. □

Back to our circle-of-friends problem: let $M = [m_{i,j}]_{1 \le i,j \le n}$ be the adjacency matrix of the Facebook friend graph, so that there are n people who belong to Facebook. Then $m_{i,j} = 1$ if and only if i and j are friends. Now let's think about

$$M^{[2]} = M * M,$$

and denote $M^{[2]} = [m_{i,j}^{[2]}]_{1 \le i,j \le n}$. Then $m_{i,j}^{[2]} = 1$ if and only if there is at least one intermediate index (call it k) such that both $m_{i,k} = 1$ and $m_{k,j} = 1$, i.e., if and only if i has a Facebook friend k, where k is also a Facebook friend of j (a "friendship chain" of length 2). Hence it should be clear that the (i, j) element of $M^{[2]}$ is nonzero if and only if there is a friendship chain of length at most two connecting Facebook members i and j.

Similarly, we define

$$M^{[3]} = M * M * M,$$
$$M^{[4]} = M * M * M * M,$$

and (more generally)

$$M^{[\ell]} = \overbrace{M \wedge M \wedge \cdots \wedge M}^{\ell \text{ times}}$$

Now denote $M^{[\ell]} = [m_{i,j}^{[\ell]}]_{1 \le i,j \le n}$. Then $m_{i,j}^{[\ell]} = 1$ if and only if there is a friendship chain of length ℓ connecting the person with label i to the person with label j. Moreover, then nonzero elements of

$$M \vee M^{[2]} \vee M^{[3]} \vee \ldots M^{[\ell]}$$

describe the pairs of Facebook friends for which there is a friendship chain of length at most ℓ.

We cannot continue this process indefinitely, simply because the set of Facebook members is finite. Since there are n people on Facebook, we see that if there is a friendship chain connecting two people, it cannot be longer than $n - 1$. (That's because the length of a friendship is one less than the number of people on the chain.) This tells us that the maximum value for ℓ in the equation above is $\ell = n - 1$. So let

$$M^* = M \vee M^{[2]} \vee M^{[3]} \vee \ldots M^{[n-1]} \tag{9.1}$$

We see that two Facebook members (labeled i and j) are connected via a friendship chain if and only if the (i, j) element of M^* is one.

The matrix M^* given by (9.1) is called the *reachability matrix* of the graph determined by M. The reason for this name is that a given entry in M^* is 1 if and only if there is a path of edges leading from vertex i to vertex j in the graph given by M, i.e., if we can reach j from i by following edges in the original graph.

Example 9.14: [SN] Suppose that the Facebook friendship matrix is

$$M = \begin{bmatrix} 0 & 1 & 0 \\ 1 & 0 & 0 \\ 0 & 1 & 0 \end{bmatrix}$$

Calculate the reachability matrix.

Solution: We have

$$M^{[2]} = \begin{bmatrix} 1 & 0 & 0 \\ 0 & 1 & 0 \\ 1 & 0 & 0 \end{bmatrix},$$

and so the reachability matrix is given by

$$M^* = M \vee M^{[2]} = \begin{bmatrix} 1 & 1 & 0 \\ 1 & 1 & 0 \\ 1 & 1 & 0 \end{bmatrix}. \qquad \square$$

Looking at the definition of M^*, we see that we compute the matrices $M^{[2]}$, $M^{[3]}$, $M^{[4]}$, ..., which are called *Boolean powers* of M. We can save ourselves a lot of work if we note that

$$\begin{aligned} M^{[2]} &= M * M = M^{[1]} * M, \\ M^{[3]} &= M * M * M = M^{[2]} * M, \\ M^{[4]} &= M * M * M * M, = M^{[3]} * M \end{aligned}$$

and (more generally)

$$M^{[\ell]} = M^{[\ell-1]} * M. \tag{9.2}$$

It's worth mentioning that the adjacency matrix is a useful representation of graphs that are large, but not huge. To be more precise, the graph must be small enough so that we can store n^2 bits in whatever computer we're using. To make this more concrete, consider the Internet graph mentioned at the beginning of this chapter. As of the writing of this text, there are close to 10^9 (a billion) hosts on the Internet, and so the adjacency matrix would use roughly 10^{18} (a quintillion) bits. Moreover, most of those bits would be zeros; we say that the graph or the matrix is *sparse*. That's because only a minuscule fraction of Internet hosts are directly connected to each other; most hosts are only indirectly connected. So an adjacency matrix would be an inefficient way to represent the Internet graph. It turns out that there *are* efficient ways to represent such large, sparse graphs (or matrices), but we will not pursue this here.

Cost of Boolean Matrix Operations

We briefly discuss the cost of computing the Boolean sum and the Boolean product of two Boolean matrices, as well as computing the reachability matrix. To keep things simple, we will only count the number of Boolean operations (in this case, $\wedge$ and $\vee$) that are performed.

Let's first discuss the simplest cases: the sum and the product.

- The Boolean sum simply uses n^2 $\vee$-operations, and nothing else. Using the O-notation introduced in Chapter 8, we see that the cost of computing the Boolean sum of two $n \times n$ matrices is $O(n^2)$.

- The Boolean product is somewhat more complicated. We need to compute n^2 elements. Each of these elements involves n $\wedge$ operations and $n-1$ $\vee$-operations. So the total number of operations needed is

  $$n^2 \times \big(n + (n-1)\big) = n^2(2n-1) = 2n^3 - n^2$$

  which is $O(n^3)$

Computing the reachability matrix M^* of a Boolean matrix M is the most complicated of all these operations. Looking at equation (9.1), we see that we need to compute $M^{[2]}, M^{[3]}, \ldots, M^{[n-1]}$, which are $n-2$ Boolean matrix powers. Each of these can be computed as one Boolean matrix product, as per (9.2). Since each matrix product has cost $O(n^3)$, the cost of this part of the computation is

$$(n-2) * O(n^3) = O(n^4).$$

Finally, we must add each of these matrix powers. Since one Boolean matrix addition can be done in $O(n^2)$ operations and there are $n-1$ matrices to add, the cost of this part of the computation is

$$(n-2) * O(n^2) = O(n^3).$$

Cost of Boolean Matrix Operations (cont.)

Combining these results, we see that we can compute M^* using $O(n^4)$ operations.

What does this $O(n^4)$ cost tell us? Suppose that parameter n increases by a factor of ten; then the running time of the algorithm increases by a factor of $10^4 = 10{,}000$. However, this is somewhat misleading, since the size n^2 of the input set has increased by a factor of 100. Since $10{,}000 = (100)^2$, the cost is 100 times the input size. More generally, we see that the $O(n^4)$ cost of this algorithm increases with the square of the input size n^2.

The obvious question here is whether there are faster algorithms for computing the reachability matrix. It turns out that faster algorithms do exist, but they are somewhat complicated, and so we will put aside such concerns for another course.

9.6 Exercises

Remark. Some of these exercises ask you to answer some questions about a graph whose picture you have been given. When looking at the picture of a weighted graph, you should always assume a weight belongs to the closest edge; if there is more than one such weight, pick the one on the left.

9.6.1 Draw the graphs that have the following vertices and edges

(a) $V = \{a, b, c\}$ and $E = \{\{a, b\}, \{b, c\}\}$.

(b) $V = \{x_1, x_2, x_3, x_4\}$ and $E = \{\{x_1, x_3\}, \{x_1, x_4\}, \{x_2, x_4\}\}$.

9.6.2 Draw the digraphs that have the following vertices and edges:

(a) $V = \{a, b, c\}$ and $E = \{(a, b), (c, b), (c, a)\}$.

(b) $V = \{a, b, c\}$ and $E = \{(a, b), (b, c), (c, a)\}$.

9.6.3 **[SN]** For each of the subgraphs of Figure A.1 in the Appendix given below in (a) through (d), answer questions (i) through (v).

(a) {Peter, Larry, Angela, Raina}

(b) {Grace, Marina, Manny, Chrissy}

(c) {Stan, Kyle, Eric}

(d) {Kip, Peter, Angela, Sandra}

(i) How many vertices are there?

(ii) How many edges are there?

(iii) Is this a directed or an undirected graph?

(iv) Is this a complete graph?

(v) Write out the set V of vertices and the set E of edges.

9.6.4 **[SN]** In a social network such as that in Figure A.1 of the Appendix, it is useful to be able to identify groups of (more than two) people who all know one another. Such a tight grouping of friends might share a lot of interests and characteristics in common. If one or two friends in such a group expresses an interest in some activity, entertainment or product, a clever social network might want to "invite" the remaining friends to discover that interest.

A group of people who all know one another is represented in a graph of the social network as a completely connected subgraph. Develop an algorithm that when given as input a vertex in a network will find a completely connected subgraph (of more than two) that contains this vertex.

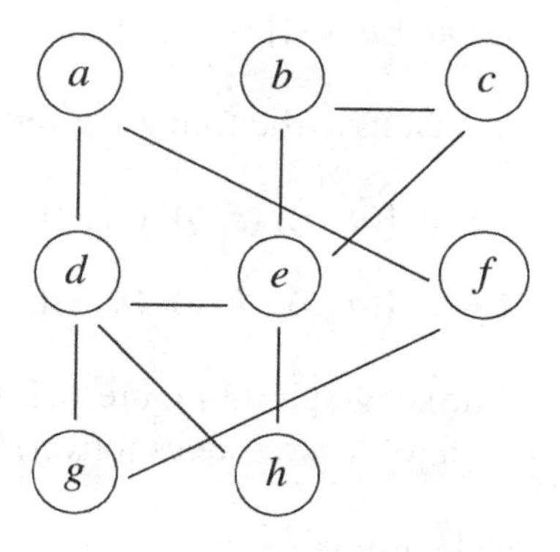

Figure 9.17: Graph for Problem 9.6.5, part 3

9.6.5 Consider each of the following graphs. In each case, determine if the graph has an Euler circuit or an Euler trail. If it has one, then show the Euler circuit or trail. If the answer is "no," then explain your reasoning.

(a) The graph in Figure 9.7 on page 290.

(b) The graph in Figure 9.9(a) on page 293.

(c) The graph in Figure 9.17.

9.6.6 Draw the graphs that have the following vertices, edges and weights.

(a) $V = \{a, b, c\}$, $E = \{(a, b), (c, b)\}$, $W = \{((a, b), 10), ((c, b), 20)\}$.

(b) $V = \{x_1, x_2, x_3, x_4\}$, $E = \{\{x_1, x_3\}, \{x_4, x_1\}, \{x_2, x_4\}\}$,
$W = \{(\{x_1, x_3\}, 55), (\{x_4, x_1\}, 1), (\{x_2, x_4\}, 44)\}$.

9.6.7 Draw a completely connected graph with four vertices, and then determine how many spanning trees you can make from this.

9.6.8 Build the MST using Prim's algorithm for the following graphs:

(a) Figure 9.11 on page 298, with edge $\{d, e\}$ deleted.

(b) Figure 9.11 on page 298, with edge $\{b, e\}$ deleted.

(c) Figure 9.10 on page 296.

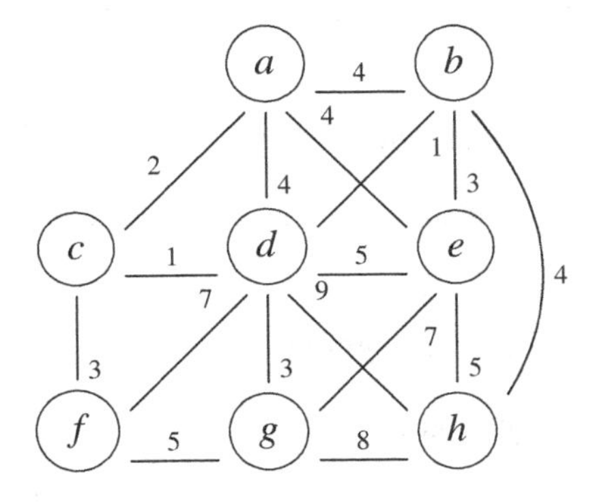

Figure 9.18: Graph for Problem 9.6.9

9.6.9 Use the graph in Figure 9.18 to answer the following questions.

(a) How many vertices does this graph have? Write the vertex set for the graph.

(b) How many edges does this graph have? Write the edge set for the graph.

(c) Does this graph have an Euler circuit? Explain why or why not.

(d) What is the weight of the graph?

(e) Draw the MST for the graph.

9.6.10 Use the graph in Figure 9.19 to answer this question.

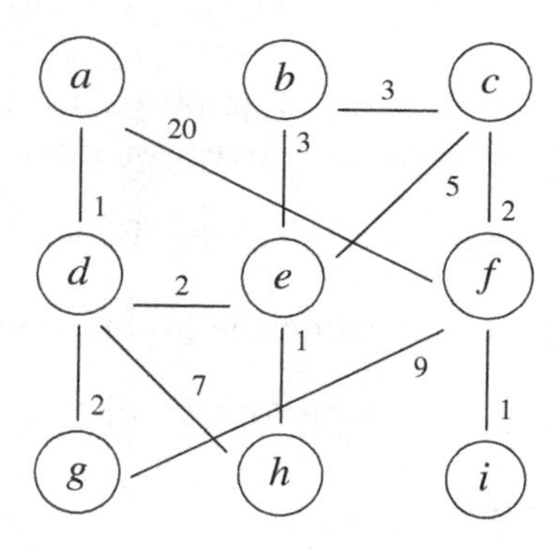

Figure 9.19: Graph for Problem 9.6.10

(a) Describe this graph as a set of vertices, edges and weights.

(b) Does this graph have an Euler circuit; show it if so, otherwise explain why not.

(c) What is the weight of this graph?

(d) Find the MST for this graph.

(e) If an edge of weight 1 is added from d to c (draw this in on the graph), how does this affect your answers for (b) and (d) above?

9.6.11 Use the graph in Figure 9.20 on page 313 to answer this question.

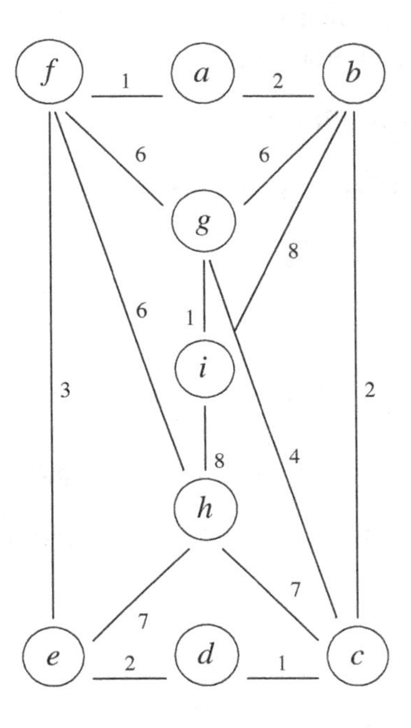

Figure 9.20: Graph for Problem 9.6.11

(a) How many vertices does this graph have? Write the vertex set for the graph.

(b) How many edges does this graph have? Write the edge set for the graph.

(c) Does this graph have an Euler circuit? Explain why or why not. If a circuit exists, list it.

(d) If an extra edge were added from vertex i to vertex e would that effect your answer? If a circuit exists, list it.

(e) What is the weight of the graph?

(f) Draw the MST for the graph.

9.6.12 Consider the graph found in Figure 9.6 on page 289.

(a) Give the adjacency matrix M for this graph. Assume that the vertices are in alphabetical order (u, r, w).

(b) What is the reachability matrix M^* for this graph?

9.6.13 Let G be the directed graph with vertex set

$$V = \{1, 2, 3, 4\}$$

and edge set

$$E = \{(1, 2), (2, 4), (3, 1), (4, 1), (4, 3)\}.$$

(a) Give the adjacency matrix M for this graph.

(b) Give the reachability matrix M for this graph.

Appendix A

Our Social Networking Example

> We lived on farms, then we lived in cities, and now we're going to live on the Internet!
>
> Sean Parker, "The Social Network"

This appendix contains the social networking example that is referenced throughout the book. We begin with Figure A.1, on page 317, which is the graphical representation of our social network first presented in the Preface on page xv. Note that each user in our network is represented as a "node" in the graph (i.e., a circle with the user's name inside it.) The solid lines that connect pairs of nodes indicate that two users have an established friendship relationship. Dashed lines with an arrowhead are used to denote that one user is requesting friendship with another (the arrowhead denotes the directionality of the friendship request). As friendship in a social networking site is a two-way relationship, directionality is not necessary and, as such, established friendship connections are shown using a solid line without arrowheads. For the purposes of the examples and exercises in this book, two people are not friends until the friendship connection is fully established.

As you study the graph you will notice that there are several completely independent groups, or clusters, of users who are currently isolated from the other users in the social network. These groups include: Stan, Kyle, and Eric; Lena, Anna, Alex, and Joanna; and poor Kip, who has no friends at all. But in these cases it is very easy to connect each cluster to the larger social network and to open up the possibility of many new friends—all it takes is one new friendship (which in the first two instances may be close to happening due to pending friendship requests). This shows the power of social networks: adding a single friend in a real world social network could open you up to potentially thousands of new friends (or business contacts).

Those of you who are members of a social network know that you provide a lot of personal information about yourself. The amount of information you provide may vary, but much of it is highly personal and collectively this information paints a very detailed picture of your life. For example, this information may specify your gender, sexual preferences, marital status, and political and religious beliefs. This information, however personal, is critical to the success of the network. Not only do social network sites rely on this information to help further your friendship connections, but they also use it to determine what products and services to market to you, thereby increasing their own profitablity.

This appendix contains three tables that provide more information about the members of the sample social network: Table A.1 provides their basic information, Table A.2 provides their education history, and Table A.3 provides their employment history. In Table A.1 we show the basic information maintained for each one of our members in our sample network. Of course this is just a small subset of the information that we typically provide in our own social networks; however, it is the most pertinent and personal. The basic information maintained in our sample network specifies the user's first name, the city they were born in, the city they currently reside in, their sex (i.e. gender), their date of birth, and their relationship status. For our purposes, we have limited relationship status to {married, single, divorced, undefined}.

Each entry in Table A.2 provides the educational history of a subset of our members. For each user represented, we show the college or university attended, the year they graduated, their degree, and their major. The entries in Table A.3 provide the employment history for a subset of our members. For each user represented we show each employer, job title, and when the job began and, if they are no longer in the position, when it ended.

The social network graph and associated tables are referenced throughout the book, mainly in the examples and exercises. This sample social network, along with the associated examples and exercises, demonstrates the relevance of the material presented in this book to our modern society.

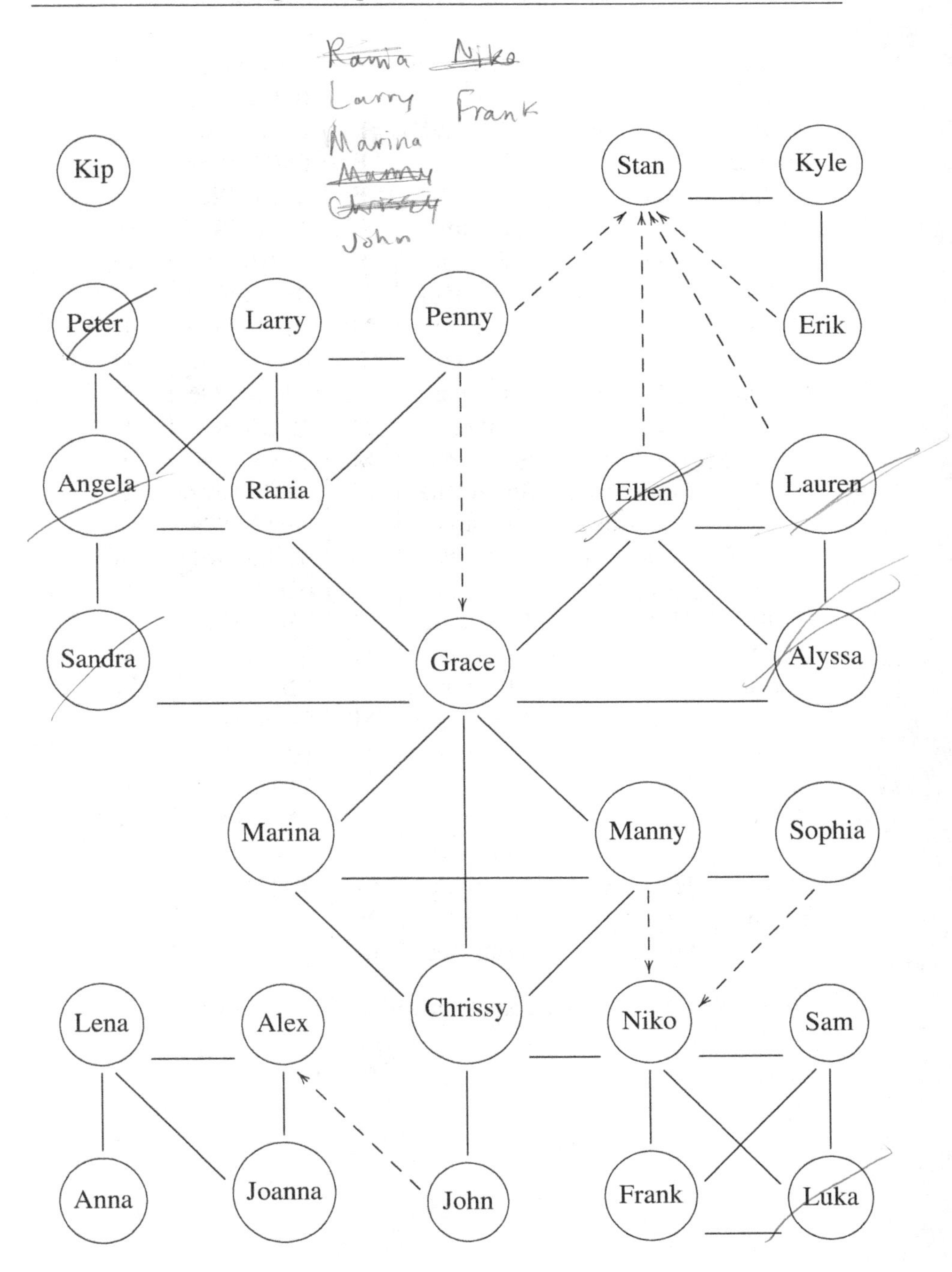

Figure A.1: A sample social network

Name	City	Hometown	Sex	Birthday	Status
Alex	Topeka	Topeka	F	02/15/1996	S
Alyssa	Hartford	Albany	F	02/01/1964	M
Angela	Charlotte	Denver	F	06/15/1967	S
Anna	Hartford	Hartford	F	5/19/1989	U
Chryssi	Boston	Boston	F	12/23/1985	S
Ellen	Hartford	Boston	F	04/01/1958	M
Erik	South Park	South Park	M	08/01/1997	S
Frank	Harrisburg	Phoenix	M	12/12/1969	D
Grace	Hartford	Boston	F	02/25/1962	U
Joanna	Topeka	Topeka	F	02/15/1996	S
John	Augusta	Atlanta	M	10/25/1991	S
Kip	South Park	South Park	M	08/01/1997	S
Kyle	South Park	South Park	M	08/01/1997	S
Larry	Albany	Boston	M	04/12/1984	M
Lauren	Hartford	Albany	F	01/12/1961	M
Lena	Topeka	Topeka	F	02/15/1996	S
Luka	Harrisburg	Columbus	M	07/29/1974	U
Manny	Boston	Boston	M	09/21/1969	M
Marina	Boston	Boston	F	12/18/1978	M
Niko	Harrisburg	Providence	M	04/08/1977	S
Penny	Boston	Boston	F	10/18/1961	U
Peter	Sacramento	Boston	M	04/18/1982	S
Rania	Atlanta	Atlanta	F	12/02/1966	D
Sandra	Juneau	Juneau	F	08/03/1967	D
Sam	Harrisburg	Nashville	M	07/12/1972	M
Sophia	Boston	Boise	F	11/25/1975	M
Stan	South Park	South Park	M	08/01/1997	S

Table A.1: Sample social network: basic information

Name	University	Class	Degree	Major
Ellen	Suffolk University	1986	JD	Criminal Law
Ellen	Harvard University	1980	BA	English
Frank	Dartmouth	1996	PhD	Physics
Frank	Dartmouth	1990	BS	Physics
Grace	Fordham University	2006	MS	Computer Science
Grace	Boston College	1984	BS	Computer Science
Larry	CUNY	2007	MBA	Finance
Larry	NYU	2005	BA	Literature
Lauren	Vassar College	1985	MA	Sociology
Lauren	Duke	1983	BA	English
Luka	MIT	2007	PhD	Computer Science
Luka	Indiana State	1997	MA	Music Theory
Luka	Ohio State	1995	BA	Classics
Marina	Tufts University	2012	PhD	Int'l Relations
Marina	Tufts University	2006	MA	Int'l Relations
Marina	Boston College	1999	BA	Politcal Science
Niko	Penn State	2007	PhD	Biochemistry
Niko	MIT	2002	BS	Biology
Sandra	Columbia	1988	BA	Journalism

Table A.2: Sample social network: education information

Name	Employer	Title	Start	End
Ellen	Barkin & Barkin	Defense Aty	1999	
Ellen	Boston DA	Prosecutor	1987	1999
Frank	Penn Medical	Physicist	1997	
Grace	Fordham University	Instructor	2006	2006
Grace	Greenville Capital	SVP	1988	1997
Larry	Goldman Sachs	Trader	2009	
Larry	Lehman Brothers	Trader	2007	2008
Luka	StringCom Inc.	CEO	2008	
Marina	Analytic Research	Researcher	2009	
Marina	ING	Analyst	1999	2003
Niko	Penn Medical	Chemist	2009	
Sandra	CNN	Sr. Correspondent	2009	
Sandra	New York Post	Sr. Reporter	1999	2009
Sandra	London Enquirer	Jr. Reporter	1989	1999

Table A.3: Sample social network: employment information

Index